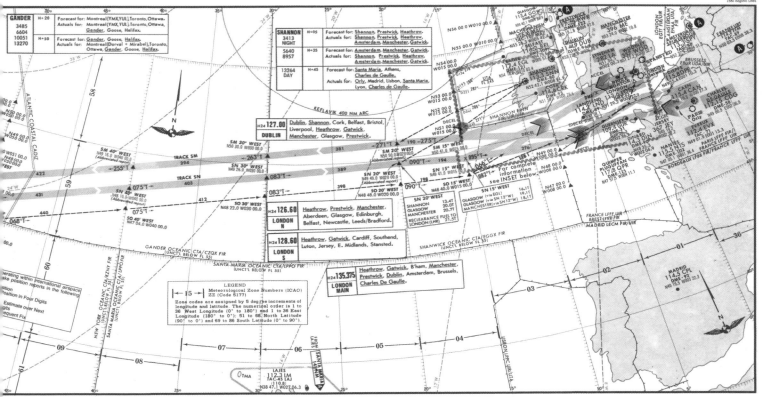

THE
CONCORDE
STORY

OSPREY
PUBLISHING

THE
CONCORDE
STORY

CHRISTOPHER ORLEBAR

'O God, who art the author of peace, and lover of concord…'
From the second Collect for Peace in the Book of Common Prayer

First published in Great Britain in 1986 by Temple Press
This 7th edition © 2011 Osprey Publishing Ltd
Midland House, West Way, Botley, Oxford, OX2 0PH, UK
44-02 23rd Street, Suite 219, Long Island City, NY 11101, USA

E-mail: info@ospreypublishing.com

ISBN 978 1 84908 163 4

Christopher Orlebar has asserted his right under the Copyright, Designs and Patents Act, 1988, to be identified as the Author of this Work.

A CIP catalogue record for this book is available from the British Library

Edited by Jasper Spencer-Smith, Tony Holmes, Shaun Barrington, Gerard Barker, Tom Milner and Charly Ford
Jacket and page layout by Myriam Bell Design, France
Index by Alan Thatcher
Typeset in Goudy Old Style and Optima
Originated by PDQ Media, Bungay, UK
Printed in China through Bookbuilders

11 12 13 14 15 10 9 8 7 6 5 4 3 2 1

Osprey Publishing is supporting the Woodland Trust, the UK's leading woodland conservation charity, by funding the dedication of trees.

www.ospreypublishing.com

Editor's note
The views, conclusions, representations and opinions in this book are those of the author. They are not those of British Airways or any of British Airways' subsidiaries.

Endpapers
These charts show Concorde's west and eastbound Atlantic tracks, 'SM' and 'SN' respectively; 'SO' is a reserve track. The acceleration and deceleration points, labelled 'ACCEL' and 'DECEL', are the points respectively where Concorde either commences acceleration from Mach 0.95, or deceleration from Mach 2. However, the deceleration point is varied a few miles either side of the position shown depending on the conditions – aircraft altitude and forecast wind – prevalent on the day.

There are a variety of routes into the United States, however the route to New York via Linnd and Sates is the one described in the chapter, 'The Flight – Deceleration'. Concorde must cross the boundary of the 'warning area' (W105 BDY) above 52,000ft when that area is 'active' with military training or test flights.

The front inset shows a plan view of the runways at Heathrow with the designations as they were when Concorde entered service in 1976: 10L, 10R, 28L and 28R. The ever-reducing magnetic variation in Britain has brought their tracks closer to 09°M and, in the opposite direction, 270°M, than to the 100°M and 280°M they used to approximate to. (°M is the measure of the direction with respect to magnetic north.) Therefore, since July 1987, they have been called: 09L, 09R, 27L, and 27R.

The back inset shows the approach path to runway 4R (the right-hand of the two north-east facing runways) at Kennedy – the runway onto which the landing was made in the chapter 'The Flight – Deceleration'.

CONTENTS

ACKNOWLEDGEMENTS

The Concorde Story was originally published in 1986 to celebrate Concorde achieving ten years in service. The sixth edition took the story of Concorde through to its final landing; but still contained ideas for future supersonic airliners that, in this age of minimising carbon dioxide emissions, simply will not happen. In this, the seventh edition, the book has been reorganised so that Concorde and the future can be viewed from the early 21st century rather than the mid-1990s. The tragedy at Gonesse is described and includes, as did the previous edition, my explanation of why the crew reacted as they did during the appalling emergency. In no way is this supposed to cast blame. Once again the reasons for Concorde's withdrawal from service are analysed. The book includes a celebration of Concorde's 27 years of service and describes how Concorde has become a most successful museum exhibit.

Since 1970, when asked for public money to fund aeronautical projects, politicians are often heard to say, 'I don't want another Concorde.' Let them not forget that Concorde was an icon. Such icons fill people with pride and more importantly enthuse young people to study the sciences and perhaps engineering. The tasks of the future, complicated by climate change and increasing population, will only be solved by well-motivated minds. If the only 'spin-off' from the Concorde project was inspirational encouragement, it was worthwhile.

I would like to thank my family and everyone else who gave me encouragement during my flying career and in the writing of this book. In particular let me thank the following:
Peter Baker, former Concorde Test Pilot, BAC
Mike Bannister, Former Chief Pilot, BA Concorde
Peter Bashford, former BAe structures
Alan Bond, MD Reaction Engines
Michael Blunt, *BA News* and Public Relations
John Britton, Chief Concorde Engineer, Airbus Industry, UK
Adam Brown, Vice-President Market Forecasts, Airbus Industrie, France
Charles Burnet, formerly of BAe Weybridge, author of *Three Centuries to Concorde* whose help and good council were invaluable

Frank Debouck, General Manager Technical and Training, Air France Concorde
Nick Edwards, Wind Tunnel Engineer, RAE Thurleigh (Bedford)
Claude Freeman, Chief Concorde Engineer BA
Alain Guilldou, Bureau Enquêtes Accidents, France
David Learmount, *Flight International*
Peter Leggett, the late and former Vice-Chancellor, Surrey University and his widow Enid
Jock Lowe, former Chief Pilot, BA
Dennis Morris, BA Engineering
Donald Pevsner, President, Concorde Spirit Tours, USA
Mike Ramsden, former Editor of *Flight International*
Gordon Roxburgh, webmaster *www.concordesst.com*
Sandy Sell, BA Engineering
Alan Simmons, Air Accident Investigation Branch, UK
Ted Talbot, former BAe Bristol, engine intakes
Donald Thompson for pointing out typographical errors in earlier editions
Brian Trubshaw, the late and former Chief Test Pilot, BAC
Brian Walpole, (former General Manager, BA Concorde) who first asked me to write the book
Ian Whittle, Pilot and aviation historian
Allan Winn, Director Brooklands Museum Trust, who acquired Concorde 202 for exhibition at Brooklands

A big thank you to all Concorde supporters; this includes those who made Concorde work and those, like me, who would rush out of the house on hearing the distinctive sound of Olympus engines just to gaze at this alliance of engineering with art. No wonder that Concorde won the title of being the Icon of the 20th century. Let its story continue to be a source of inspiration for the future.

Christopher Orlebar
Oxshott, August 2010

DEDICATION

In remembrance of all those who died in the tragedy at Gonesse near Paris on 25 July 2000: the three flight crew, six cabin crew and 100 passengers on Concorde F-BTSC and the four people on the ground.

FOREWORD TO THE SEVENTH EDITION

By Lord Marshall of Knightsbridge

Lord Marshall led British Airways for 21 years, initially as Chief Executive (1983–1993), then as Chairman and Chief Executive and then, from 1996, as non-executive Chairman. He retired from the airline in July 2004.

There have been few more powerful symbols of all-round excellence than Concorde. The power, speed, beauty and sheer airborne grace of this truly unique aircraft have thrilled the masses around the world for more than 30 years. Over its commercial career, the British Airways Concorde fleet carried more than 2.5 million passengers at supersonic speed on regular scheduled services and many special flights around all parts of the globe. Millions more people came to cherish and admire Concorde for the style it displayed and the technological endeavour it stood for. Quite simply, Concorde lifted our spirits.

The decision to retire the supersonic fleet at the end of October 2003, after almost 28 years of service and 140 million miles, was prompted by compelling technical and economic reasons. Even so, it was tough to make and hard to announce. At British Airways, we resolved to create a special 'finale' programme to ensure that Concorde would bow out in triumph and with dignity. We believe this was achieved.

We also wanted to ensure that the Concorde fleet would not just fade away, but go to places of honour at some of the great aviation museums and other special locations in Britain and around the world. It was important to us that, although the aircraft would no longer be flying, Concorde would live on as an enduring symbol of aerospace excellence.

Accompanying Concorde for much of its career has been Christopher Orlebar's much-loved book, *The Concorde Story* and I am delighted that this new edition will play its special part in helping us to keep the Concorde experience alive and relevant.

All of us at British Airways are immensely proud to have had Concorde as our supersonic flagship for so many momentous years. It has been an honour and a privilege to be one of only two airlines capable of making commercial supersonic flight a breathtaking reality.

When a supersonic, or even hypersonic, successor to Concorde might emerge is, at the time of writing, unknown. What we can be sure of, is that aviation in this century can be relied on to come up with developments every bit as advanced and exciting as Concorde was in the last.

In the meantime, let us continue to enjoy and marvel at *The Concorde Story*, so eloquently told by one who knows this remarkable aircraft well.

Lord Marshall of Knightsbridge
Chairman, British Airways Plc

ABOUT SUPERSONIC FLIGHT

This book is about the conquest of flight at speeds faster than sound. It covers the synthesis of the host of disciplines which were brought together to build and operate the most superlative form of transport yet built by man – Concorde. Yet Concorde exists in the ordinary physical world. There was no divine intervention which altered the laws of nature to favour Concorde. Its success has been achieved by the devotion of countless people who have been inspired by the concept. It ranks with the architectural marvels of the world – the Pyramids, the Gothic Cathedrals and the Taj Mahal.

This book was originally written to celebrate the tenth anniversary of Concorde commercial operations, 1976–1986. Now well into the 21st anniversary it has been revised and updated. What follows is a brief description of some of the technical terms and concepts that the more inquisitive reader may find useful. In some instances these are expanded in the chapters that follow.

Ernst Mach (1838–1916), an Austrian physicist, observed that airflows obeyed different laws as they approached the speed of sound. In recognition of his work, the term 'Mach number' was named after him. The Mach number of a moving body is the ratio of its speed to that of the speed of sound in the fluid in which it is travelling. The fluid in which aircraft travel is, of course, the air in the atmosphere. When an aircraft's speed is the same as the speed of sound it is said to have a Mach number of one – Mach 1.

Ernst Mach was a positivist who argued strongly against the notion that matter was made up of atoms –

The airspeed indicator on Concorde shows just over 500kts; whilst the Machmeter indicates just less than Mach 2, this corresponds to a true airspeed of 1,150kts. The apparent discrepancy is due to the low density of the air found at 53,000ft causing the aircraft to 'feel' only 500kts. See 'The Flight Envelope' in the Appendices.

they could not be seen. The explanation of why the airflow obeyed different laws as the speed of sound is approached is much more easily understood on the assumption that atoms and molecules do exist. It is therefore curious that Mach did not agree with what has now become the accepted truth.

Air is a gas which is made up of molecules moving randomly at high speed. The molecules' speed varies with temperature. They move more quickly when it is hot than when it is cold. The speed of sound in air is the speed at which vibrations or changes of pressure are transmitted through the air. As this speed is a function of the speed the molecules move, sound travels quicker in hot air than in cold.

At 15°C (59°F), the average temperature found at sea level, the speed of sound is about 760mph (660 knots). Between 50,000 and 60,000ft, where the temperature averages –57°C (–71°F), it is 660mph (573 knots). It varies in fact with the square root of absolute temperature (absolute zero, or zero degrees Kelvin, is –273°C [–459.4°F] thus the freezing point of water is 273°K and its boiling point 373°K).

An aircraft in flight can be travelling subsonically, i.e. below the speed of sound, or supersonically, i.e. above the speed of sound. The airflow, as Mach observed, is governed by two quite different laws depending on whether the aircraft is flying at subsonic or supersonic speeds. There is, however, a third regime – the 'transonic'. It covers a speed range from roughly Mach 0.75 to Mach 1.3. It is called 'transonic' because both supersonic and subsonic (governed by their respective laws) are present simultaneously.

At first sight it is not obvious why a supersonic airflow can be present when an aircraft is flying at less than Mach 1. The reason is that an aircraft in flight separates the air. The airflow over the more bulbous parts of the aircraft (over a protruding cockpit or over the top of the wings) is accelerated during this deflection; it now has an increased kinetic energy. The total energy of the airflow must remain constant

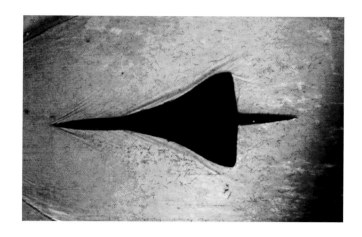

The shock waves formed in the air around a supersonic aircraft are analogous to the waves formed by a moving boat. Here water is flowing over a thin silhouette of Concorde to demonstrate the analogy. Although waves on the surface of water are formed in two dimensions, the shock waves on an aircraft propagate in three dimensions, forming a cone at the nose.

(otherwise something would be being given for nothing). So what it gains in kinetic energy it must lose in potential, in this case pressure energy. This is consistent with the law of energy conservation, so repeating: as the 'moving' energy of the air increases, its pressure energy must reduce. Low pressure on the top surface of a wing sucks it upwards, or gives it 'lift' which is just what is wanted. However when the speed of an aircraft exceeds about Mach 0.75 the faster moving air over the wings can reach Mach 1. The Mach number of the onset of transonic flight varies from aircraft to aircraft and is dependant on wing shape, wing sweep and fuselage profile.

When the aircraft flies faster than Mach 1 some of the air, depending on the shape of the aeroplane, is pulled along by the aircraft, so that some of it flows subsonically over the surface. Usually by the time the aircraft has reached about Mach 1.3 all the airflow over it is supersonic. Flight at transonic speeds was to prove more difficult than flight at supersonic speeds.

As an aircraft flies it experiences resistance from the air. This is known as 'drag'. The amount of lift compared to drag that an aircraft experiences at a given

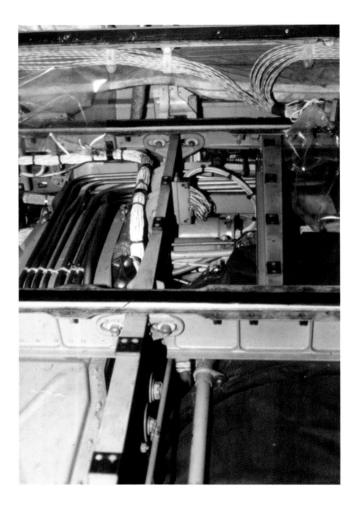

Kinetic heating causes the fuselage to lengthen – 20cms in Concorde's case. To prevent the cabin floor from distorting, the seat rail supports have elongated slots, here one on either side of the cross-beam.

speed is a measure of its aerodynamic efficiency. This is known as the lift to drag ratio, commonly written as L/D. Drag is caused by the shape of the aircraft (form drag), by the surface of the aircraft (skin friction drag) and by the work the wings have to do to give lift (induced drag). The first two increase with speed, while induced drag decreases with speed.

At speeds below about 300 knots (1 knot = 1.15mph), the airflow over an aircraft behaves like an incompressible fluid (like water). Above 300 knots the molecules of air, having had less 'warning' of the impending disturbance, become compressed. The 'warning' travels at the speed of sound. As the aircraft travels closer to Mach 1 the molecules have progressively less time in which to redistribute themselves before the passage of the aircraft. Now they build up and make, at Mach 1, a distinct 'compression wave' or 'shock wave'. This wave, analogous to a bow wave of a boat, increases the drag. Hence the notion that there was some kind of barrier at Mach 1, although projectiles like bullets had exceeded the speed of sound since the early 1700s. At the rear of an aircraft flying above Mach 1 is another compression wave where the air reverts to its original condition, having steadily reduced to a lower pressure during the passage of the aircraft. These two shock waves can usually be heard on the ground as a double boom.

At first sight it is difficult to visualise that the air ahead of an object moving at subsonic speed starts to redistribute itself before the arrival of the object. Observation of light airborne objects, like snowflakes or thistle seeds, becoming wafted over the windscreen of a moving car demonstrates the effect. Heavier objects like grit and raindrops do not deflect. Should the car accelerate towards Mach 1, the deflection of even the light objects would not happen. Their first 'knowledge' of the car would be the car itself.

Once the sound barrier had been 'broken' another barrier appeared – the heat barrier. The faster an aircraft flies the more it is heated by the air which surrounds it. This is not an effect which is suddenly manifest, as it is present at all speeds, but does not become significant until the aircraft is travelling supersonically. It is caused by the compression of the air as it is accelerated during the passage of the aircraft. When any gas is compressed its temperature rises, as is readily observed when pumping up a bicycle tyre. The end of the pump becomes very hot due to the compression of the air within it. On an aircraft this effect is known as kinetic heating.

The temperature rise in degrees centigrade can be calculated approximately by dividing the true airspeed of the aircraft (in mph) by 100 and then squaring. At Mach 2, about 1,320mph, the rise is 174°C (345°F); at Mach 3 it is 392°C (737°F). These figures are for

An F/A-18 Hornet from USS *Carl Vinson* performs a supersonic fly-by. The condensation shows the low pressure surrounding the fuselage. At the tail the pressure returns rapidly to ambient so the cloud disappears. (US Navy)

the point experiencing the greatest temperature rise, usually the tip of the aircraft's nose, where the air is accelerated to the speed of the aircraft. To this figure must be added the static air temperature. Above 37,000ft this averages −57°C (−71°F), varying from place to place over the globe. At Mach 2 the final top temperature is thus 117°C (243°F) and at Mach 3 it is 335°C (635°F). Kinetic heating has an immense bearing on the choice of cruising speed for a supersonic transport aircraft (SST) and the materials from which it will be built. Aluminium alloys can easily cope with temperatures up to about 130°C (266°F); higher than that, more expensive titanium alloys must be used.

To monitor the temperature experienced by the nose of fast moving aircraft there is a gauge in the cockpit. It simply reads degrees centigrade. There are other instruments like the altimeter and compass which say, with a fair degree of accuracy, how high the aircraft is and in which direction it is pointing. There is one instrument, however, which to the untutored eye appears as if it is constantly telling a frightful lie (unless it is observed at sea level) and that is the airspeed indicator.

At sea level the speed shown on the airspeed indicator corresponds quite closely to the true airspeed (TAS). But as the aircraft climbs into the thinner air the indicator registers an airspeed which is less than true. At Mach 2 and 55,000ft the true airspeed would be 1,150 knots, but because of the low density of the air, the indicated airspeed is only 500 knots.

Aircraft are flown with special reference to indicated airspeed since this is a measure of what the aircraft 'feels'. Knowledge of what the aircraft 'feels' tells the pilots how the aircraft can perform; how close it is, for instance, to losing the ability to give lift (stalling). Or, in the case of Concorde, how much extra thrust will be required to overcome the high induced drag when flying at landing speeds.

Indicated airspeed also shows the pilot how great an angle of attack the aircraft is experiencing. The angle of attack is the angle between the aircraft and the flow of air. As the speed falls so this angle increases, thus lift is maintained. If constant altitude is to be maintained during deceleration, the aircraft must be pitched up to compensate for the increased angle of attack. When descending to land, Concorde has quite a high pitch attitude to the horizontal (11 degrees). This is because of the high angle of attack (14 degrees) associated with low indicated airspeed. The difference between angle of attack and pitch attitude gives the angle of descent or climb – 3 degrees being the usual slope during an approach to land.

In the next chapter we shall see how the theory of supersonic flight was learnt and then how cleverly it was applied in the design of Concorde.

THE BACKGROUND

'Pull steadily, watch the altimeter, don't flap and don't expect anything much to happen until below 15,000 to 20,000ft…' This advice was contained in a document written by George Bulman, Chief Test Pilot of Hawker Aircraft in 1943. It was intended for pilots who found themselves in aircraft diving at speeds close to that of sound.

A number of Allied fighter aircraft had experienced strange effects when diving at such speeds during the Second World War. Not the least of these effects had been the dangerous tendency for the controls to do the reverse of what was expected of them and for the whole aircraft to buffet and shake, sometimes to the point of structural failure. To preserve frontline fighter pilots and to investigate flight at speeds close to Mach 1 – thus forestalling any German advance in the new science – a series of high speed dives was planned.

The aircraft chosen for these investigations was the Photo Reconnaissance Spitfire (PR Mark XI) fitted with the Merlin 61 engine. Such a Spitfire was capable of climbing to 40,000ft – high even by the standard of the subsonic jets of the 21st century. To achieve the highest speeds the power dive had to be started from as high as possible. For that and the reason of its relatively thin wings, the Spitfire PR Mark XI was considered a suitable aircraft for research into flight at transonic speed.

Some time in 1943, Squadron Leader Martindale, a test pilot for Rolls-Royce in civilian life, found himself in a 45 degree dive from 40,000ft. After 36 seconds from the start of the dive, descending through 29,000ft,

In 1943, when Squadron Leader Martindale was performing a series of near-sonic dives in a Spitfire, the best way of recording the changes of air pressure at the various points around the aircraft was by filming the gauges attached to a bulkhead – actually two bulkheads and two cine cameras.

The Gloster E28/39, Britain's first jet-propelled aircraft, which first flew on 14 May 1941. Its Whittle-designed, Power Jets engine gave a modest 850lb of thrust. The engine planned for the US SST was to have produced close on 70,000lb of thrust. The big fan engines in service since the mid-1990s are capable of delivering well over 100,000lb of thrust.

the Spitfire attained Mach 0.9. From about Mach 0.75 upwards some of the airflow became compressed and therefore subject to different aerodynamic laws. The Spitfire experienced increased drag, loss of lift and a tendency to pitch further nose down. Uncorrected the nose-down pitch would increase the dive angle. To prevent this, the pilot would pull the stick back to apply 'up elevator'. At transonic speeds such action exposed the aircraft to another risk. Due to lack of stiffness in the tailplane and the great force of a compressed airstream on the up-going elevator, the leading edge of the tailplane could bend upwards leaving the elevator trailing in the slipstream. This resulted in the controls

achieving the reverse effect of what was expected, namely pitching the aircraft further nose down. This effect is quite separate from the nose dropping due to the redistribution of the lift.

Entering the warmer atmosphere the Spitfire found itself in an environment where the speed of sound was greater. Although still travelling at roughly the same true airspeed, its speed with respect to the speed of sound (Mach number) was lower. The danger now was that the laws governing the airflow at the slower speeds would suddenly be restored. The 'up elevator' which had maintained the angle of descent could now, at the smaller Mach numbers, reassert itself with unnerving suddenness, pulling the aircraft out of the dive at such a rate that the wings might fold upwards, or flutter like a flag in a stiff breeze. There were no ejection seats in those days.

These investigations into flight at speeds close to Mach 1 would have been given immense publicity in peacetime. In the event they were cloaked in wartime secrecy and by the time the story was told, it appeared rather pedestrian compared to the rocket- and jet-propelled attempts on the sound barrier during the late 1940s. It is of lasting tribute to the Spitfire that it behaved better at such speeds than some of the later aircraft of supposedly more advanced aerodynamic design.

The Rolls-Royce Merlin 61 engine, which powered the PR XI, had been developed and fitted to the Spitfire Mark IX in response to a formidable new fighter fielded by the Germans in 1941: the Focke-Wulf 190. The increased performance of the Merlin 61 was largely due to the effectiveness of its supercharger. This added 70mph to the top speed and 10,000ft to the Spitfire's maximum altitude. In consequence it out-performed the Focke-Wulf 190.

A supercharger is a device that compresses the air before it undergoes further compression in the cylinders of a piston engine. The supercharger on the

Merlin 61 engine had been refined to near perfection by a brilliant young engineer at Rolls-Royce in Derby – Stanley Hooker.

Stanley Hooker (later Sir Stanley Hooker) had been recruited in 1938 by the works manager of Rolls-Royce Ernest W. Hives (later Lord Hives of Duffield) with the somewhat enigmatic words: 'You are not much of an engineer, are you?'! Hooker's work on superchargers was soon to have great relevance in the development of the jet engine, since the centrifugal compressors of the early

The Me 163 Komet, Germany's rocket-propelled interceptor, the design of which was based on research by Dr Alexander Lippisch, the 'father' of tailless and delta-winged aircraft. The aircraft on the left is a prototype, lacking the nose-mounted generator of the production aircraft on the right. The Me 163's rocket propulsion was available for eight to ten minutes, giving speeds in excess of 550mph. Its swept wings allowed flight at high subsonic Mach numbers before the transonic drag rise took effect. The planform of the DH108 was somewhat similar to this aircraft. (Courtesy EN Archive)

German wind tunnel technology was the best in the world. Airflows well in excess of Mach 1 could be achieved. The V2 rocket was supersonic, hence the need to build such a device. This compressor was shipped to the Royal Aircraft Establishment (RAE) Thurleigh Bedford at the end of the Second World War. On the compressor the name of the oil pump manufacturer name is visible, evidence of its German origin. This wind tunnel was dismantled c. 2006.

jet engines were similar in principle to, although much larger than, the supercharger of the Merlin. The jet engine turned out to be the most suitable power plant for continuous supersonic flight. Eleven years after his work on the Merlin supercharger, Stanley Hooker began the transformation of the Bristol Olympus engine, destined, in its most superlative version, to power Concorde.

The 'barrier' to greater speed from the Spitfire during its transonic dive was due to two effects. One was the large increase of drag it experienced as the compression waves built up at Mach 0.9; the other was the loss of thrust experienced by the propeller. The combined effect of aircraft forward speed and the turning speed of the propeller ensured that the outer portions of the propeller blades were travelling at supersonic speeds. At such speeds the propeller becomes very inefficient. Thus the Spitfire lost thrust just as its drag increased. A rocket-propelled, rather than a gravity-assisted, Spitfire might have been able to maintain Mach 0.9 for longer.

Rocket propulsion has the advantage over other kinds of propulsion in that the thrust it generates is not dependent on the forward speed of the aircraft. Nor is its thrust dependent on altitude, if anything it increases as the pressure of air around it decreases. Although rocket-propelled aircraft are spared the need of having any kind of air intakes, they suffer the greater burden of having to carry extra fuel weight in the form of liquid oxygen (or some chemical suitable for combustion with the fuel). They therefore do not give propulsion suitable for sustained flight.

Happily by 1943 there had appeared an engine which looked capable of giving sustained thrust both at subsonic and supersonic speeds. This was the jet engine. This new engine gave thrust not through a propeller, but by virtue of its high speed jet efflux. The fact that its intake might be in a supersonic airflow would be, if anything, a bonus, since the incoming air could be

slowed down and therefore 'supercharged' prior to entering the compressor on the engine, making the engine yet more efficient.

The inventor of the jet engine principle, Sir Frank Whittle, said that he was regarded at the time as a crazy optimist. One of his early proposals (in 1935), envisaged an engine giving 111lb of thrust to a 2,000lb aircraft travelling at 500mph at an altitude of 69,000ft. It is anyone's guess what people would have said had he predicted that a 350,000lb aircraft (including the fuel) would be capable of travelling at 1,320mph at 55,000ft with each of its four engines giving, at that altitude, 10,000lb of thrust.

Jet-propelled flight became a reality in Britain in 1941 with the first flight of the Gloster E28/39. A Whittle-designed engine, the Power Jets W2/700, was chosen, in 1943, as the power plant for Britain's first supersonic project: the Miles M52 (E24/43). It was hoped optimistically that the M52 would exceed 1,000mph (about Mach 1.5) in level flight. There was a great deal of controversy over the choice of wing section for the aircraft. Thin wings, suitable for supersonic flight, seemed to be the obvious choice, but their lack of efficiency at subsonic speeds meant that the aircraft would not be able to climb to such a high altitude as one fitted with conventional wings; the higher the altitude, the higher the Mach number that could be attained in the ensuing dive. Once comfortably supersonic it was hoped that the aircraft could level off maintaining speed – engine thrust having increased to cope with the higher drag associated with supersonic flight.

Above: The large wind tunnel at RAE Thurleigh Bedford was capable of producing an airflow of Mach 2.5 at 4 atmospheres.

Left: The wind tunnel at RAE Thurleigh Bedford. The higher pressure allows a more accurate assessment to be made of a scaled-down model with respect to the full sized aircraft. The 'sting' is the point where the wind tunnel model is attached. In this 1962 photograph a model, looking very much like how Concorde turned out, is under test. (Photo by SSPL/Getty Images)

The Vampire was Britain's second jet fighter. The DH108 was effectively a Vampire with swept wings and no tail booms. Once the advantage of swept wings was perceived, a Vampire was modified, ending up looking like the Komet – but with jet rather than rocket propulsion

The W2/700 was one of the last engines built by the Power Jets company. Under the leadership of Frank Whittle, a small and dedicated team working in appallingly primitive surroundings near Rugby in England (Sir Stanley Hooker likened Whittle's study to a rabbit hutch) had researched, designed and built a viable jet engine. After an unsatisfactory period of liaison with the Rover company, Rolls-Royce became involved with Power Jets and, under the leadership of Stanley Hooker, set about developing and producing jet engines in quantity.

The lineage of the modern jet engine owes far more to the early British engines than to the jet engines developed in Germany. On observing the immense length of runway required by the Me 262 (the German jet-powered fighter) on take-off, an onlooker remarked to Frank Whittle, who was also present, that it was no wonder that Hitler had wanted to extend the Third Reich.

Soon it became evident that more research was needed to give sufficient thrust to the M52. The reason that the Ministry of Supply gave for the cancellation of the project in 1946, however, was that they were concerned for the safety of its pilot. In 1948, a pilotless, rocket-propelled, scaled down version of the M52 did achieve about Mach 1.4, which proved that its planform with straight wings could cope with supersonic flight. However, at the end of the Second World War, British design was influenced by swept wings which gave better performance at both supersonic and high subsonic speeds.

Before the outbreak of the Second World War a German scientist, Professor Adolf Büsemann, had published findings suggesting that swept wings would perform better than straight wings in supersonic flight. One significant reason for this is that the shock waves generated at the nose of an aircraft miss the wing tips if these are swept back. Thus there is not the added complication of interference between wings and shock waves. Astonishingly these findings, freely available to all, went unnoticed.

Equally remarkable was the failure of the Allies to understand why the wings were swept on the rocket-propelled German fighter (the Messerschmitt Me 163 Komet). The Komet was a tailless swept-winged interceptor, which was very fast, achieving Mach 0.8 (about 550mph) for short periods. The Allies believed that its speed could be ascribed to its being rocket-propelled. The reason for the swept wings was thought to be so that adequate control could be retained in pitch (the control of the angle of the fuselage with respect to the horizontal). Though both beliefs were partly correct, what was not understood was that swept wings allowed the aircraft to fly at speeds closer to Mach 1, since the increase in drag was delayed until these higher speeds were reached.

From the description of the Spitfire tests, it will be remembered that some of the air becomes supersonic over an aircraft even though that aircraft is travelling subsonically. It is the portion of the supersonic airflow which causes the drag rise.

Thick wings experience this transonic drag rise at lower Mach numbers than thin wings. Wings must support the weight of the aircraft and have within them some space to contain fuel. High speed aircraft need both strength and as much space for fuel as possible. The conflict can be resolved by sweeping the wings either forward or back. Although the ratio of the thickness of the wing to the distance from leading to trailing edge remains constant, the sweepback makes this ratio 'appear' smaller to the airflow. Now the air is less disturbed by the passage of the aircraft, so the aircraft can fly faster before the compression waves start to apply their drag increase. The benefit of this discovery, made by Professor Albert Betz (a German

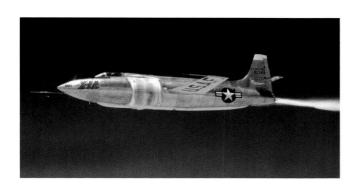

Left: A Bell X-1A, a later version of the X-1 aircraft in which Chuck Yeager became the first pilot to exceed the speed of sound in October 1947. It had straight but thin wings and a tailplane situated out of the slipstream from the wing. (USAF)

Below: The DH108, Britain's first transonic swept-wing aircraft. Control in roll and pitch on aircraft without tailplanes has to be achieved through the surfaces on the trailing edge of the wing. John Derry exceeded the speed of sound in this aircraft (VW 120) – the first British aircraft to do so.

aerodynamicist), has been applied to jet-propelled airliners, virtually all of which have swept back wings.

To research flight with a swept-winged aircraft, the de Havilland company modified a Vampire, a single engine, straight-winged fighter, with its tail and fin surfaces connected via twin booms to the wings. The booms and the tail surface were removed, swept back wings fitted, and a swept fin was placed over the rear end of the fuselage. It resembled the Komet in many outward respects. The DH108 was partly intended as a research aircraft for what became the Comet airliner. At that time the philosophy among several British aircraft manufacturers had been that it was desirable to perfect a 'flying wing' to reduce the aerodynamic drag and the extra weight caused by tailplane surfaces. Sadly, after the loss of several such aircraft to lack of control in pitch, this philosophy had to be abandoned.

'Flying wings', as tailless aircraft were dubbed, would have had to carry their payload within the wing. For people to stand inside the wing it would have had to have been at least 7ft thick. The wing span would have been huge to cope with such a thickness. Furthermore without a tailplane there was no way of aerodynamically balancing wing flaps. Without flaps either the landing speed would have had to have been very fast or the wing too big for efficient cruise. It would thus have been an uneconomic proposition – the de Havilland Comet had a tail!

Two other British aircraft design philosophies had their origins in the immediate post-war era. One was to produce a wing capable of sustaining laminar flow. Hitherto the flow of air over a wing had always been impeded by that little turbulent layer of air close to the wing known as the boundary layer. Laminar flow control promised a very marked reduction in aerodynamic drag.

Below: The de Havilland Comet 1 represented Britain's great leap forward in the post-war aviation era. Stresses at the right angular corners of the windows were higher than predicted; hence the metal fatigue that led to fatal crashes. But for the accidents de Havilland might have retained their world lead in jet airliner construction. (© Museum of Flight/Corbis)

The de Havilland Comet 4. The windows are now oval-shaped following experience with the Comet 1. The engines are buried at the wing roots – a philosophy preferred in Britain to the underslung pods chosen by Boeing and other US manufacturers. (© Museum of Flight/Corbis)

The other philosophy was the notion that the aerodynamic drag associated with the engines and intakes could be reduced by placing the engine air intakes at the wing roots feeding air to engines that were 'buried', as on the Comet 4, rather than in pods, as on the Boeing 707. Placing the engines at the rear of the aircraft, as with the VC10, may have given the wing the advantage in lift at all speeds; but came at the cost of needing a stronger, and therefore heavier, wing root than that of the Boeing 707 whose engine weight was borne by the wing but outboard of the root. This layout of the 1950s has become the norm as exemplified by the Airbus 380 and Boeing 787.

The laminar flow concept failed largely because it seemed impossible to retain a completely clean wing.

Even a squashed insect, acquired during take-off, upset the laminar flow. Removal of the boundary layerthrough minute holes drilled in the surface of the wing was another solution, though rather impractical, at least in those days. However, laminar flow control was put forward again in the mid-1980s as a means of reducing drag on an Advanced Supersonic Transport (AST).

The presence of a boundary layer can be readily observed on a car windscreen during rain. Even when driving at 70mph the smaller drops already settled on the glass surface remain stationary. The larger ones which protrude out of the boundary layer are moved by the airstream but not that quickly. It is some millimetres away from the screen that the speed of the airstream assumes its expected value.

In 1948, John Derry, Chief Test Pilot of de Havilland, became, more by accident than by design, the first pilot in Britain to exceed Mach 1, in a DH108 during a test flight investigating transonic flight. Whilst

Dassault Mystère IV. This swept-wing French fighter-bomber was in production by 1955, and it provided useful data about flight at transonic speeds.

attempting to recover from a dive at Mach 0.97, his aeroplane exceeded the vertical and probably achieved Mach 1.02. He was lucky to survive. Geoffrey de Havilland, son of the founder of the firm, had been killed in a DH108 almost exactly two years earlier probably having experienced similar very severe oscillations in pitch.

By the early 1950s data had been acquired which would make the next steps in supersonic flight slightly easier. The first aim then was to build a jet fighter capable of intercepting the mammoth Russian nuclear bombers which were under development, and were capable of cruising between 30,000 and 40,000ft at around Mach 0.8.

The fighter had to have an excellent rate of climb and at least double the bomber's speed for a successful interception long before the bomber could reach its target. The second aim was to build a supersonic bomber which, because of its speed, would itself be very difficult to intercept.

Long-range supersonic flight, however, was going to be much more difficult to achieve than the short duration dash required of an interceptor. But should a bomber become a reality then a transport derivative could follow, such was the pattern generally established during the 50 years of aircraft development. But as far as the development of an SST was concerned, the introduction of missiles even capable of intercepting supersonic bombers at high altitude was to upset this evolutionary pattern. The SST would have to be developed almost in its own right.

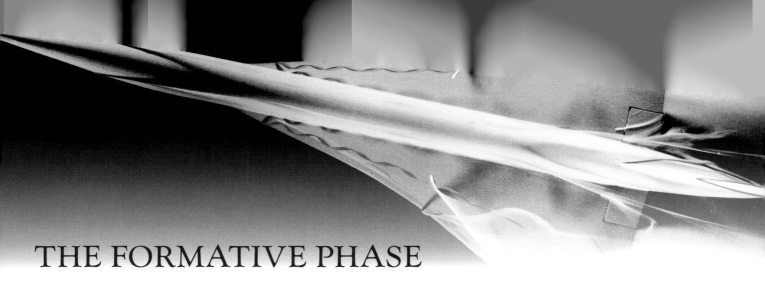

THE FORMATIVE PHASE

Prompted by the preponderance of designs for supersonic interceptor aircraft that appeared in the early 1950s people began to wonder whether a civil supersonic transport (SST) could one day be a possibility. With London to New York flight times being of the order of 18 hours and more, especially against headwinds and with refuelling stops, the idea of doing the same journey in a fifth of the time must have seemed preposterous.

Nevertheless Sir Arnold Hall, director of the Royal Aircraft Establishment (RAE) Farnborough in the early 1950s, asked Morien Morgan, already involved with the Advanced Fighter Project Group and one of the greatest proponents for civil supersonic flight, to chair a small committee to look into the possibility of building a civil supersonic transport. The reported findings, in 1954, suggested that it might just be possible to fly 15 passengers from London to New York at Mach 2 in an aircraft with an all up weight (maximum take-off weight including the passengers and the fuel) of 300,000lb (136,000kg). For comparison Concorde carried 100 passengers (up to 128 with reduced space between seat rows) over that range with an all up weight of 408,000lb (185,000kg). The envisaged 15-seater SST was based on the design of the Avro 730 supersonic bomber project that was cancelled in 1957. It had thin unswept wings with engines mounted on the wing tips. But for an SST to be economically viable, a better aerodynamic shape would have to be devised.

Such a shape did appear with dramatic impact, in the form of the British Fairey Delta 2. On 10 March 1956 Peter Twiss (Chief Test Pilot of Fairey Aviation) exceeded the previous world airspeed record, held by the American Colonel Haynes in an F-100 Super Sabre, by the handsome margin of 310mph. A reheated Rolls-Royce Avon RA5 turbojet propelled the FD2 at the astonishing speed of 1,132mph (Mach 1.7).

In November 1956 the Supersonic Transport Aircraft Committee (STAC) was formed, chaired by Morien Morgan. Hearing evidence from 17 interested bodies, comprising the aircraft industry, airlines and ministries, its conclusions were reported in 1959. The result was Concorde, which made its maiden flight ten years later.

Contributors to its success were the FD2 and the P1, both offsprings of Morgan's Advanced Fighter Project Group which had been formed in 1948. The success of these two aircraft banished the notion of some kind of intractable barrier at Mach 1, the speed of sound. It is interesting to look at the different design philosophies.

Both designers produced a different planform: the P1 with swept wings and a tailplane, while the FD2 a delta wing without a tail. Tailless designs had encountered many problems, as exemplified by Geoffrey de Havilland's experience in the DH108. Therefore English Electric wanted to ensure stability and control through use of a tail surface, especially for flight in the tricky transonic range. Fairey believed,

The Fairey Delta 2 (FD2). Fairey believed that the delta wing with no tailplane was the optimum solution for a supersonic aircraft. Peter Twiss achieved a new world air speed record of 1,132mph in the FD2 in March 1956. An FD2 (BAC 221) was modified with wings to the 'Ogee' planform to research separated airflow.

however, that a 60 degree delta with powerful trailing edge control surfaces would provide a stable aircraft throughout this range. After all, experience had pointed to the fact that tail surfaces were a major source of aerodynamic buffet at transonic speeds – why run this risk by having a tail at all?

As it turned out both designs were supremely successful. In different ways both contributed to Concorde's eventual success: the FD2, because it was converted into the modified delta shape that was to be applied to Concorde; the P1, because a great deal was

learned about the variable exhaust nozzle on the Rolls-Royce Avon engine, a necessary device for the efficient use of a variable thrust reheat system. It is the variable exhaust nozzle system on the Olympus engine, as fitted to Concorde, which eked out every last remaining ounce of thrust and fuel economy from the jet efflux both with and without reheat.

'Reheat', or 'afterburner' as it is known in the United States, is a device which can be fitted to the rear of the main jet engine, within the exhaust duct, to give an extra 'push' to the aircraft. It consists of a ring of nozzles which spray fuel to be burned in the engine exhaust gases, which still have sufficient oxygen to support combustion, prior to its leaving through the exhaust nozzle. As the velocity of the jet gases are so much greater with reheat on, the exhaust nozzle must vary in area to accommodate two situations. On the FD2 the reheat was either on or off, thus the exhaust nozzle only had the two positions – open or wide open. On the P1 the reheat was variable so the exhaust nozzle had to be variable as well. On the Olympus 593, as fitted to Concorde, not only is there a variable nozzle to accommodate the reheat, but there is a secondary nozzle which, amongst other things, forms a divergent duct for efficiency during the supersonic cruise. Furthermore the primary nozzle area varies continually, even without reheat, thus keeping the Olympus engine constantly in tune (see Appendices).

The STAC met regularly from November 1956 until March 1959. Among its recommendations was one which strongly favoured building a long-range SST to carry 150 passengers from London to New York non-stop with a cruising speed of not less than Mach 1.8. The committee regarded cruising speeds approaching Mach 3 as feasible, but technically difficult due to the heating effect discussed in the first

chapter ('About Supersonic Flight'). At Mach 3 the highest temperature experienced by the fuselage is around 335°C (635°F) thus not only is it much more difficult to keep the occupants cool, but the skin has to be made from the more expensive alloys of titanium and perhaps even from stainless steel. The cheaper and well tested aluminium alloy, although highly satisfactory at temperatures up to 130°C (266°F), does not cope with temperatures of over 300°C (572°F) at all. A Mach 3 version was regarded by the STAC as a possible second generation SST. There was also a recommendation for the construction of a shorter-range SST to cruise at Mach 1.2.

Between London and New York a cruising speed of Mach 3, compared to Mach 2, would reduce the total journey time by about 40 minutes, about 20%. The real gain of a Mach 2 airliner was that it could more than double the speed of the current subsonic airliners, which were designed to cruise at just above Mach 0.8. Any economic gains associated with the yet quicker journey times made possible at a speed of Mach 3 would be more than wiped out by the fact that a Mach 3 SST would cost disproportionately more to

The Bristol 221 was just the other FD2; but modified to test the stability of the aircraft when lifted by vortices alone.

Sir Barnes Wallis with a model of Swallow. Although the cockpit and undercarriage have been extended for this photograph, the wings are in the fully swept position for supersonic flight – a combination that would not have been found in practice. Surrounding Sir Barnes are examples of his earlier engineering efforts including bombs, airships, the Wellington bomber.

with a longitudinal swelling in which would be housed the passengers – virtually a slender delta flying wing.

Duncan Sandys' somewhat infamous defence White Paper in 1957 had put a temporary halt to the development of manned military aircraft in Britain (including the Avro 730 supersonic bomber) in favour of missiles. Sandys had been a Royal Artillery Officer during the Second World War, which may have influenced his judgement. So a rather stunned aircraft industry must have welcomed the opportunity to research and possibly build an SST. By 1959, when Duncan Sandys had commissioned the feasibility studies for an SST, plans for the new British Supersonic Tactical Strike Reconnaissance Aircraft – the TSR2 – had been unveiled. However, in 1966 Britain's Labour government cancelled the TSR2 as well as other military aircraft projects. To quote Sir Stanley Hooker on British military aviation policy: 'We were in, we were out; then we were in again and finally out. It was more like a boat race than a policy.'

Among other designs which were also put forward at about this time was the Swallow designed by Sir Barnes Wallis, an aircraft in which variable sweep had been taken to the extreme. The engines, mounted outboard on the wings, were pivoted and could be used for stability and control in place of aerodynamic surfaces. Furthermore their movement with the wing sweep contributed to the balance of the aircraft. Aerodynamic balance was maintained on Concorde by moving the fuel to change the centre of gravity.

An 'arrow' planform is attractive because it is very efficient aerodynamically at supersonic speeds. The ability to vary the sweep of the wings (variable geometry) could maximise this advantage. In the unswept configuration, the engines on Swallow were well outboard of the centre line. Following an engine failure there might have been a difficulty in keeping directional control; especially as there was no fin and rudder. In 1969 I attended a lecture given by Barnes Wallis at the Gatwick Manor Hotel. When asked about the absence of

build. Development costs were put at £75–£95 million for the Mach 1.8 SST and £50–£80 million for the slower, shorter-range versions, at 1959 values. It was considered that they would cost about 50% more to operate than a similarly sized subsonic aircraft.

Following the recommendations of the STAC, feasibility studies were commissioned by the Ministry of Defence, then headed by Duncan Sandys. Bristol produced studies of an SST with a conventional fuselage and slender delta wings, but Hawker Siddeley provided a study for an integral layout – a slender delta a fin, he said with a twinkle that he had a solution which he had not divulged to Boeing. Boeing had just given up its variable sweep supersonic transport design (the Boeing 2707-200) in favour of a delta layout.

A 'laterally' thought out way of how to achieve variable geometry in flight is to 'slew' a single wing in its entirety. The wing could either be mounted on a conventional fuselage or, as studied by Handley Page (HP), without. In 1961 HP proposed an SST carrying its payload of 150 passengers within the wing. The slew would be varied according to Mach number. Effectively the passengers would be facing in the direction of flight at low speed then sideways during the supersonic cruise. Without digital computers the flight control laws might have been impossible to apply. Even had it worked,

Concorde model in a hydrodynamic tunnel. The coloured fluids indicate the vortex flow over the wings; this type of 'separated' flow over a slender delta gives excellent lift, but with increasing drag as the speed decreases.

marketing an aircraft which lacked the elegance of symmetry could have proved wearisome, not helped by it being called a *slew* wing.

A third proposal envisaged a slender delta with a large number of lifting engines to give it a vertical take-off capability. No fewer than 50 such engines were thought to be necessary on a transatlantic SST! The rate of fuel burn of the lifting engines was expected to be very high – and the result of them all failing to start simultaneously prior to landing would have left the aircraft in a somewhat dangerous state.

These three proposals (and the Boeing 2707-200) failed to exploit the discovery that extra lift could be obtained by allowing the airflow to separate over the upper surface of a slender delta wing. To give lift, under normal circumstances, the flow of air over a wing must not separate from the wing. The airflow travels slightly further over the top surface than under the bottom surface and so has to move more quickly over the top

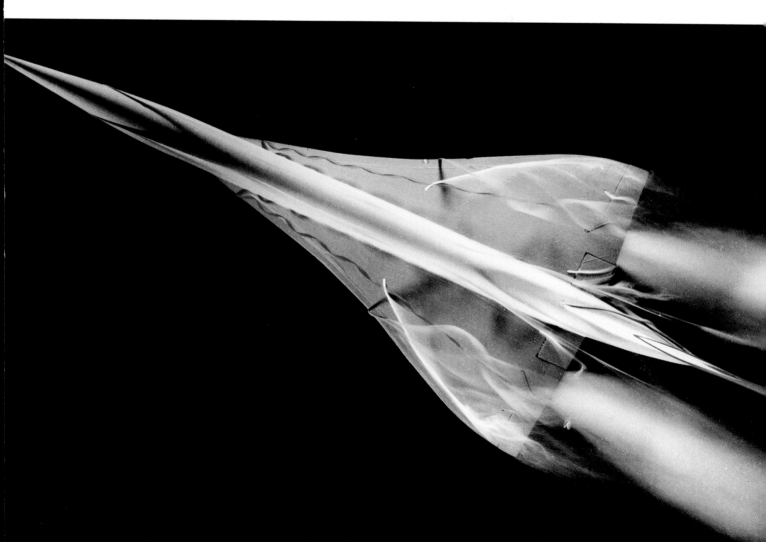

than the bottom. As it accelerates its pressure drops. In this way the wing is 'sucked' upwards, rendering lift to it and to whatever the wing is attached. Should this airflow become turbulent, thus breaking away from its ordered flow, the lift will suddenly reduce; the wing is then said to 'stall'. At the tip of a wing in normal flight is a small vortex – a whirlpool of air rotating, in this case, about a horizontal axis; it is caused by the higher pressure air underneath twisting into the lower pressure air on top. Such a vortex, on a larger scale, can be harnessed to give lift to a slender delta wing in slow speed flight; high lift devices (flaps and slats) then are not necessary.

A big contribution in the study of flow separation was made by EC Maskell of RAE. He and the late Dr Dietrich Küchemann, head of Aerodynamics at RAE (who had been a scientist in Germany during the Second World War, later 'inherited' by Britain,)

Rooms full of draughtsmen have been overtaken by the digital computer. Concorde just preceded the change. Perhaps years would have been shorn off the 14-year gestation period between starting the project and its entry into service.

laid the foundation of this new technique. The slender delta was already performing well in the transonic and suspersonic regimes, could it also perform at low speeds? A practical test was needed to prove both that the theory was correct and that no dangerous side effects would prevent it being applied to an SST.

Consequently the feasibility programme that was commissioned following the final report from the STAC, included a contract to Handley Page to build an aircraft to test what the flying paper darts, beloved by schoolboys, had suggested: that vortex lift could give sufficiently stable lift to allow a slender delta flying at low speed to be controllable.

Here the low speed characteristics of Concorde are being investigated in a wind tunnel. As the speed decays the lift is maintained by increasing the angle of attack which strengthens the vortices. This increases the drag which eventually exceeds full thrust. The zero rate of climb speed varies with weight and thrust available (number of engines running). Although Concorde does not stall in the conventional manner it has an anti-stall system. At 13.5 degrees angle of attack (AoA) the 'auto stabilisation' system starts to pitch the nose down. At 16.5 degrees AoA the 'stick' (control column) is shaken and the autopilot disconnects. If 19 degrees AoA is reached, the elevons pitch the aircraft 4 degrees down and the stick is wobbled by the hydraulic jack that normally gives 'artificial feel'. Concorde was designed so that pilots converting to it from subsonic airliners would find it easy.

The resulting aircraft was the experimental Handley Page 115. It was powered by a single Viper jet engine. Although its 76 degree swept wings gave the illusion that it was built for great speed it was designed purely to explore flight at low speeds – it did not even have a retractable undercarriage.

The handling of the HP 115 was better than anyone had dared to hope. It gave excellent lift at slow speed. Even landing in a cross-wind, a situation which might have caused difficulty, was trouble-free.

Success with the HP 115 proved that the resulting SST would not have to be complicated with flaps and slats to enable its wing to give sufficient lift at take-off and landing speeds. If it had one small vice it was that as the speed became slower the drag caused by the wings giving lift became greater. Although this obviates the need for air-brakes, the extra drag has to be overcome

Overleaf: The HP 115 in flight near Farnborough. The slender delta wing was fitted with detachable leading edges, allowing various aerodynamic shapes to be tested. This aircraft proved that a separated airflow could give lift at slow speeds. First flown in August 1961, its success paved the way for Concorde by showing that complex high lift devices for slow speed flight would not be necessary for slender deltas.

by increasing the thrust. Provided there is plenty of thrust available, all is well; but the danger lay in the pilot not spotting the speed loss and thus not applying the extra thrust required by the lower speed. Without proper monitoring of the speed it could fall off to the point, not of the wings being unable to give lift and so stalling, but to the point where there was not enough thrust to overcome the extra drag associated with the low speed. To overcome this problem it was recognised that such an aircraft would have to be fitted with an automatic throttle. Indeed on Concorde there was just such a device and it worked extremely well too. However, Concorde was surprisingly easy to fly in the very rare event of the auto-throttle not being available.

Although the slender delta promised to have a better lift to drag ratio (L/D) at supersonic speeds than other planforms, the unpalatable fact remains that there is a drag rise as Mach 1 is exceeded. In fact the L/D dropped, on Concorde, from about 12 at Mach 0.95 to 7.5 at Mach 2. How then was the future SST going to have enough range to cross the Atlantic?

Happily jet-engine efficiency can be designed to improve with speed, more than compensating for the loss of aerodynamic efficiency. The projected SST was expected therefore to travel further (or at least as far) on a gallon of fuel at its supersonic cruising speed as it

Left: Sir Morien Morgan first considered an SST in 1948, he chaired the STAC and became a tireless campaigner for the building of what became Concorde. He was a Welshman and along with other Welsh aeronautical engineers, notably Handel Davies, was regarded as being part of the 'Taffia'. Sadly he died in 1978, but he did see Concorde start services.

Below: The engine intakes on Concorde. The two moveable 'ramps' can be clearly seen inside the top of the box. The extent of the movement on the forward one (labelled 'DO NOT USE …') can be clearly seen on the vertical dividing wall between each intake and its neighbour, in the form of 'rubbing' marks. Correct positioning of the ramps to focus the shock wave on the intake's lower lip was fundamental to engine efficiency in supersonic flight. During subsonic flight the ramps are raised, as shown here. See Appendices.

could at its subsonic cruising speed, Mach 0.95. This fortunate state of affairs is brought about because the efficiency of the turbojet engine is very much enhanced if the air is pre-compressed prior to engine entry. At speeds in excess of Mach 1.4 sufficient pre-compression can be arranged to occur inside the intake system where the air is slowed – hence compressed – to about half the speed of sound before it reaches the engine. Nobody doubted that the intake system would need some kind of variable 'throat' so that it could cope efficiently with various speeds and especially those in excess of Mach 1.5. Since engine efficiency improves with speed up to Mach 3, there was a great incentive to go as fast as the kinetic heating of the structure would allow – Mach 2.2 appeared the most attractive speed if a well-tried and tested aluminium alloy was used.

However, Mach 2.04, which reduced the maximum temperature from 156°C to 127°C (313°F to 261°F), was ultimately chosen since the lower temperature gave less of a problem with the oil and fuel.

Left: Dietrich Küchemann CBE was a German aeronautical engineer. At the end of the Second World War he came to the RAE at Farnborough. His studies of the slender delta lead to the shape of Concorde. In particular he recognised that vortex lift would solve the problem of how to make the slender delta fly sufficiently slowly for take-off and landing.

Below: The slewed-wing proposal from Handley Page. Runways would have had to have taken on different proportions if this 1961 idea had taken root!

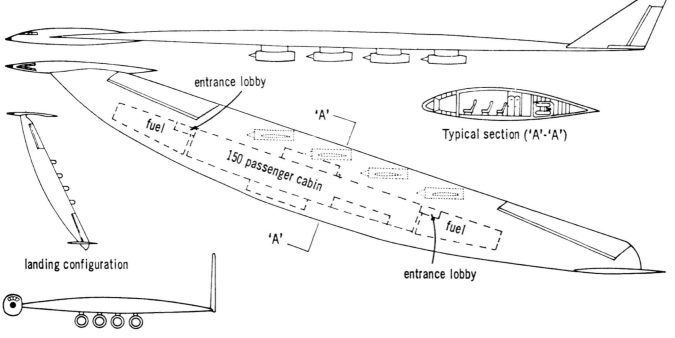

This artist's impression was never likely to work. All fifty-plus engines would need to be started together using fuel at a high rate from tanks which should have been where the engines are shown. The advantage of supersonic flight over short distances was very slight, so why is it shown in British European Airways livery?

By the late 1950s, it was realised, tantalisingly, that aerodynamic and thermodynamic development had progressed far enough to allow an SST to become a reality. The question now became: would disappointments with the performance of British civil aviation spur government and industry into turning the theory of a paper dart into a practical SST?

British civil aviation in the years after the Second World War had hoped to equal or better anything the Americans could produce. It was true that two major projects had been abandoned – the Princess Flying Boat and the Bristol Brabazon. But on the plus side the pure jet de Havilland Comet 1 and the turbo-prop Vickers Viscount were about to appear. At Bristol another potential world beater, the turbo-prop Britannia, was also taking shape.

The Britannia, it was hoped, would replace most long-range piston-propelled airliners. The engine chosen to power the Britannia was the Proteus. Originally designed to fit within the wing of the Brabazon, it had an air intake arrangement which caused the air to turn through 180 degrees before entering the engine. At 20,000ft in the moist air found over the tropics ice formed at the 180 degree elbow causing the engine to flame out (stop) on an early Britannia. Fears of a similar, if not worse, problem over the Atlantic brought matters to a head. However, the moisture content over the Atlantic at 20,000ft is not nearly as great as it is over the tropics. According to Sir Stanley Hooker a relatively easy modification would have sufficed. Nevertheless for worldwide

operations a total cure had to be found. In the end a system of igniter plugs solved the problem and BOAC (British Overseas Airway Corporation, later BA) accepted the Britannia – some two years late – and just in time to be overtaken by the jets on the North Atlantic, notably the Boeing 707. Instead of 180 only 80 Britannias were sold, furthermore a jet-powered version of the Britannia was beyond the resources of Bristol. The Britannia did serve with the Royal Air Force in the transport role, and many of the ex-BOAC aircraft saw service with holiday charter and freight airlines throughout the world.

On 15 May 1952 the era of the passenger was opened when a de Havilland Comet 1 set off for Johannesburg from Heathrow. To extol the virtue of vibration-free flight, a picture was published of a coin balanced vertically on its rim on a table in the cabin of a Comet 1 whilst in the cruise at an altitude in excess of 30,000ft. The days of the piston propeller driven airliner were numbered. This 'number' would have been even less had disaster not struck the Comet 1. Following a searching investigation, the reason for the tragic loss of two Comet 1 airliners in early 1954 was revealed. Cracks originating in the corner of a window in the pressurised fuselage for a radio navigation aerial had propagated so quickly that the cabin shell had just exploded. There followed an intense period of redesign. In October 1958, two weeks before the Boeing 707, the Comet 4 inaugurated the

Above, left: The buckets or secondary nozzles on Concorde. The left-hand engine shows the nozzle in its divergent position as it would be in supersonic flight. The right-hand engine shows the nozzle as it would be for take-off, the gap allowing ambient air to break up the boundary to the jet efflux making it less noisy.

Left: Before the main landing gear can be retracted it has to be shortened so that it can fit into the undercarriage bays. The 'chines' – aerodynamic surfaces on either side of the nose – are there to reduce side slip. A vortex forms on the upstream surface which generates a force which pulls the aircraft straight. Without these the fin would have been taller to avoid being blanketed during flight at high angles of attack.

first transatlantic jet passenger service. The windows on the latest Comet had rounded corners; but notably the US aircraft industry became privy to the lessons too, the same expensive mistake would not be made over there. The world lead in airliner manufacture, so nearly won by the Comet, had been lost. Would that lead, aeronautical sages wondered, be restored if an airliner capable of another halving of flight times were built?

In the meantime the British civil airliner programme was in a muddle. It had been galling for Sir George Edwards, Chairman of Vickers, that in 1956 the British Government had allowed BOAC to purchase the Boeing 707 from America. In the previous year, BOAC and the RAF had turned down the British equivalent to the 707 – the Vickers VC7 (V-1000 in its military form). Based on the Vickers Valiant bomber, it had engines buried in the inboard section of the wing with intakes at the wing roots. The VC7 was scrapped, to be replaced by the Vickers VC10.

The Vickers VC10, with its clean wing and four rear-mounted engines, was originally conceived as an airliner capable of operating from high-altitude airfields on hot days over the old 'Empire Routes'. Later its range was increased to compete with the Boeing 707. Although it was beloved by passengers as soon as it appeared in 1962, it was too late to sell in large numbers – the Boeing 707 had got in first. It typified the frightful divergence in the relationship between the British Government, the nationalised carriers (BOAC and

P1 Lightning. The tail surface can be clearly seen. The English Electric designers considered that inclusion of a tail would give the stability and control required for supersonic flight provided it was positioned away from the wake of the wing. The exhaust nozzles of the two Avon engines are mounted one above the other. The nozzles could vary the area of the exit orifice – wide open with full reheat in use, and diminishing with intermediate amounts of reheat. Perfection of variable jet nozzles was essential before a commercial SST could become a reality.

The Bristol Britannia. The big British hope of the 1950s. Icing problems on the Bristol Proteus engines delayed its entry into service. Too soon it was overtaken by the long-range jet airliners.

BEA) and the aircraft industry. The Government, according to Sir Basil Smallpeice (Managing Director of BOAC from 1956–1963), had expected BOAC to support the British aircraft industry by buying British aircraft it did not want and then having to bear some of the financial burden incurred during the final stages of the development of the new aircraft. At the same time BOAC was expected to pay the Government a fixed rate of interest on the capital borrowed. It was galling to BOAC that a large quantity of that capital, in 1957, had had to be spent on acquiring ten DC7Cs to fill the gap caused by the late delivery of the Britannia 312.

Thus BOAC regarded the VC10 with a great deal of caution, fearing that the Government would force them to have a large number of VC10s they simply did not want. In the end some kind of compromise was reached, but not without acrimony.

Forewarned is forearmed. Sir Basil Smallpeice and Sir Matthew Slattery of BOAC made it very clear to the

Opposite: The Vicker VC10 was the last large passenger aircraft built in Britain. Seen here taking off from Brooklands climbing steeply, it had excellent runway performance, designed as it was for the 'Empire Routes' – the hot climate, high altitude airfields of Africa. Had it not had such performance it might never have left Brooklands where it was built. It had equally good landing qualities. Note the banked track of the former Brooklands motor racing circuit as 'Standard' VC10 G-ARVK lands.

A Concorde model painted in BOAC livery. By the time it made its first commercial flight, BOAC had merged with BEA to become British Airways. The VC10s in the background at Heathrow represent the last large all-British airliner, and barely 50 were built.

British Government in 1960 that although they would be delighted to assist with any future SST, this would be at no financial risk to the airline and no orders would be placed unless the SST made economic sense. In the end the Government agreed to underwrite that risk.

It is true that there had been one British success – the short-haul Vickers Viscount, with sales exceeding 400. However, by the late 1950s and early 1960s, in the words of Sir Stanley Hooker, 'We've lost in the civil market to the Americans. Now these Boeing 707s and DC8s cannot possibly last for more than about ten years. Therefore in order to collar our fair share we must look towards building a supersonic airliner with a European partner.'

AGREEMENT WITH FRANCE

By 1961, Britain had convinced herself that it was both highly desirable and technically possible to build an Atlantic-range (3,700 statute miles) SST. There was one piece of the jigsaw missing, however; could a suitable partner be found to share the large cost of such a project? The right choice of partner, it was hoped, would lead not only to a successful SST but to co-operation in a host of other fields. A European partner, for instance, might pave the way for Britain's entry into the Common Market. At that time Russia and America were pouring millions into the exploration of space, without, apart from technical advance, any prospect of financial return. An SST promised technical advance and a financial return, the latter largely on the grounds that speed had always attracted customers.

Partnership with an American company was considered. But the Americans were more interested in Mach 3 designs. Such speeds had already been ruled out by the British designers on account of the kinetic heating problem.

The only European country with the ability to undertake such a project in the early 1960s was France. The nation had to rebuild its once proud aircraft industry from the ashes of the Second World War. She was, in 1960, one of the few nations – with Russia, the United States, Sweden, Britain and Canada – to have designed, built and tested supersonic aircraft. Nevertheless the French aircraft industry would benefit enormously from Britain's technical know-how which had been enhanced rather than destroyed by the Second World War. As it turned out, Britain was to benefit from France's determination to see the project through.

Starting with the straight-winged Trident, first flown in 1953, France had built a series of supersonic research aircraft. It was the Trident which on one occasion caused some concern in the echelons of the British Radar Warning System. Charles Burnet (BAe Weybridge) who witnessed the incident said: 'One evening in the late 1950s a worried controller phoned Boscombe Down to find out if there were any test aircraft airborne from there at that time. Apparently a target had been detected flying down the Channel from the east at over 1,000mph – it turned out to be the French Trident!' This aircraft was powered by two wing tip mounted turbojets and a rocket unit in the fuselage.

After this came the delta-winged Gerfault which was capable, in 1954, of exceeding Mach 1 by use of a turbojet alone – without reheat or rocket assistance. But most spectacular of the French research aircraft was the Nord Griffon. It had a propulsion unit which consisted of a turbojet within a huge outer casing. This outer casing formed a ramjet. The turbojet engine was used for slow speed flight and for starting the ramjet. The Griffon achieved Mach 1.85 (about 1,216mph) in 1957.

Ramjets work on the principle that due to the high forward speed, usually above Mach 1.5, there is sufficient natural compression of the air to supplant the

The Mirage III. The French pursued the delta planform after the British success with the FD2. Developed in the late 1950s and sold worldwide, the Mirage has been a great success for the French aircraft industry. It is here seen in Australian service. (USAF)

rotating compressor of a turbojet. No compressor means no need for a turbine to drive the compressor hence no moving parts. Air entering the intake becomes compressed and is then heated by the burning fuel, giving thrust to the engine as it accelerates out of the propelling nozzle. A ramjet does not give efficient propulsion until speeds of around Mach 4 are reached.

Many missiles are propelled by ramjets, but only once they have achieved a suitable speed, usually gained from a rocket engine. Concorde travelled too slowly to benefit from ramjet propulsion.

On the civil side the French had some success with the subsonic twin-engined Sud-Est Caravelle airliner. Its nose section was in fact the same as the British de Havilland Comet's. There was a lot of collaboration.

Incidentally Concorde was to inherit, via this lineage, the triple hydraulic system labelled blue, green and, as a backup, yellow, which was originally fitted to the Comet.

Concurrent with the Nord Griffon were the swept-winged Mysteres and the delta-winged Mirage fighters. The next logical step for Sud Aviation, as it had become by 1961, was to investigate the possibility of building an SST. Sud Aviation had been formed from Sud-Est, run by George Hereil, the original proposer of the Caravelle, and Sud Ouest.

Accordingly, at the Paris Air Show of June 1961 there appeared a model of a slender delta SST called the Super Caravelle. Although never built, it was designed to carry a payload of 70 passengers over 2,000 miles.

The Super Caravelle was thus not to have transatlantic range. In fact there was to be some wrangling between the British and French over the question of range. The British, in spite of a mild flirtation with an 'M-winged' medium-range Mach 1.2 aircraft, had never considered it worthwhile to build anything with less range than London (or Paris) to New York. The Super Caravelle bore a remarkable resemblance to the BAC 223, the latest proposal from Archibald Russell, Technical Director of Bristol Aircraft which, by late 1961, had become part of the British Aircraft Corporation (BAC). Included in the merger were Bristol Aeroplane Co Ltd, English Electric Aviation Ltd, builders of the P1 Lightning, Hunting

Right: Sir George Edwards was Chairman of the British Aircraft Corporation during the important stages of Concorde's development. During his career he worked with Barnes Wallis on bouncing bombs, when he applied some technology learnt on the cricket pitch. This was to apply backspin to the bomb to help it roll close to the dam to achieve maximum effect.

Below: The medium-range M-winged Mach 1.2 proposal from Armstrong Whitworth. The cone at the nose would have kept the nose shock wave clear of the wings. Area ruling, apparent where the fuselage is waisted, was intended to keep the drag down. Concorde was not quite long enough to have required area ruling.

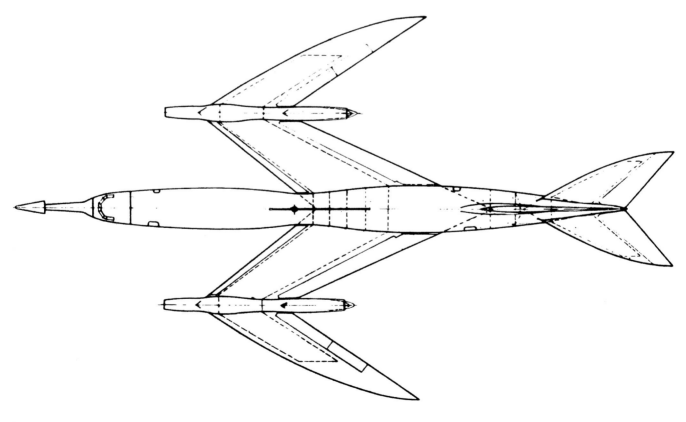

The Rolls-Royce/SNECMA Olympus 593-610 engine, powerplant for Concorde. Between 50,000 and 60,000ft at Mach 2 it is the world's most efficient aero engine, with more than 40% thermal efficiency. With reheat on take-off it delivers over 38,000lb of thrust, during the supersonic cruise around 10,000lb.

Aircraft Ltd and Vickers-Armstrong (Aircraft) Ltd, builders of the VC10.

The BAC 223 study had followed a proposal for a larger version – the Bristol Type 198 which had been considered too ambitious to gain Ministry approval.

The Type 198 design had been originally submitted as one of the feasibility studies which had followed the recommendations of the STAC. With transatlantic range at a cruise speed of Mach 2.2, it was to have had six Olympus engines in a bank on top of the wing under the fin. The projected all up weight was to be 385,000lb, with a passenger load of 132. Concorde turned out to be only 4% heavier with a similar load carrying ability and four Olympus engines of increased thrust instead of the original six.

The BAC 223 on the other hand, although still with transatlantic range, had a proposed weight of 270,000lb and was to be powered by four under-wing mounted Olympus engines. This question of size as well as range was already causing much controversy. A great deal of unnecessary expense was incurred building tools and jigs for designs that were not produced.

Duncan Sandys, perceiving the political and economic advantages of collaboration between France and Britain, had encouraged Sir George Edwards, Chairman of the new BAC, to explore the possibilities of co-operation with his French opposite number,

Georges Hereil of Sud Aviation. At first, there did not appear to be much of a basis for co-operation due to the differing range philosophies. So discussions centred around collaboration on a common engine, on systems, electrics and hydraulics.

To achieve better coordination, the French Minister of Transport and Britain's Minister of Aviation met in September 1961. This resulted in a firm directive to the participating firms to resolve their differences. But collaboration proved more difficult for the airframe manufacturers, whose ideas on range were becoming more entrenched, than for the engine manufacturers. There was only one engine that could sensibly be used for both the medium- and long-haul SST: the Bristol-Siddeley Olympus.

Bristol Aero Engines had persevered privately with the Olympus in spite of Government policy which had decreed, in 1957, that there would be just the one large engine programme, centred on the Rolls-Royce Conway

engine. Had it not been for the wisdom and foresight of Sir Reginald Verdon-Smith, Chairman of the Bristol Aeroplane Company, there would have been no engine immediately available and suitable for an SST.

There followed more meetings between the respective ministers to encourage collaboration. There was little result at first, other than the making of a few conciliatory gestures – the building of the tail section

---

Concorde 02 – the French-constructed pre-production aircraft F-WTSA displaying both makers' names against an Alpine backdrop. By the time this picture was taken Sud Aviation had become Aerospatiale. Each country was to operate the aircraft that it had constructed. Note the extended tail cone compared to the prototypes and British pre-production Concorde 01. Every minor improvement was vital in order to achieve the design range and payload of at least Paris to Washington. The greater the fineness ratio (length to maximum diameter) the better the lift to drag ratio (about 7½ on Concorde at Mach 2, although 10 or 11 is desirable in a successor). Keeping fineness ratio high on a design with a shorter fuselage compromises internal volume.

of the Super VC10 was subcontracted out to a Sud Aviation subsidiary.

Then in 1962 de Gaulle came to power in France. General Andre Puget replaced Georges Hereil, while in Britain Julian Amery became the Minister of Aviation. These changes propelled the two countries into an agreement with one another. De Gaulle saw this as a way of improving the French aircraft industry while Puget, unlike Hereil, was on the same wavelength as George Edwards; Amery was to seal this concord with an unbreakable treaty.

Just before the signing of the treaty, the Americans became perturbed by the progress being made in the field of SSTs in Europe. Accordingly, Eugene Black, ex-Chairman of the World Bank and a prominent and influential individual in the US corridors of power, attempted to dissuade Julian Amery from continuing with his plans. According to Geoffrey Knight (one-time director of BAC) in his book *Concorde: the Inside Story* this merely had the effect of encouraging Amery to go ahead. The Americans were afraid that Britain's gamble was going to pay off. It turned out that Eugene Black was about to become Chairman

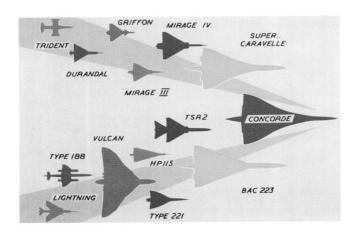

Left: Concorde's family tree. The Super Caravelle and BAC 223 were only design studies. The Type 221 was a modification of the FD2. The Bristol 188 was built out of stainless steel; but it failed to reach a Mach where this would have been an advantage in avoiding the effects of kinetic heating. However much was learnt from it 'on how not to build a supersonic aircraft' according to Ted Talbot, a Bristol Aeronautical Engineer involved in both the Bristol 188 and Concorde projects.

Below: The distribution of responsibilities for the manufacture of Concorde. Broadly the breakdown between France and Britain was 60:40 for the airframe and 40:60 for the engine.

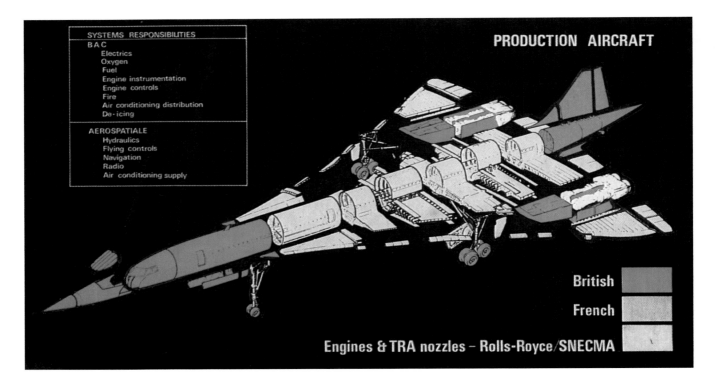

of a high-powered committee set up to study the possibility of the United States building an SST.

On 29 November 1962 an agreement between the governments of France and Britain was signed and registered at The Hague. The agreement comprised seven articles. It included clauses about each country having equal responsibility for the project, bearing equal shares of the cost and sharing equally the proceeds of sales. Two aircraft – a long-range and a medium-range one – were envisaged. Every effort was to be made to carry out this programme with equal attention to both the medium and long-range versions. There were to be two integrated organisations taken

The Concorde production line at Toulouse, France; the other line was at Filton, England. It is doubtful that any future collaborative project on the scale of Concorde would duplicate production lines. Development continued even after the 'production' aircraft were being built. Thanks to a US 'Change of Requirement', 22 titanium straps (two between frames 9–20 – roughly between the front door and the leading edge of the wing) had to be incorporated. These are fixed internally within the top of the forward part of the fuselage from floor level on one side to the other. The straps were 3 inches wide but only 12 thousandths of an inch thick. This modification was retrofitted, at no little expense, to the earlier Concordes but included in the later versions. Every aircraft has to be equipped with a means of passenger escape and, for over water flights, life rafts. On Concorde the four forward exits are equipped with inflatable 'slide/rafts' – a slide that can be converted into a raft. This picture shows the extent to which the overwing exit has to open in order to accommodate the slide/raft box fixed to its lower half, here resting on the wing.

According to some commentators the cost of developing Concorde was increased by one-third due to its being a collaborative project. There were two production lines – one at Bristol (Filton) in England, and the other at Toulouse in France. Duplication of jigs and tools were two elements in this process, whilst transporting pieces made in France to Britain and vice-versa was another. Here the 'Guppy' is being loaded with a British-built tailfin and tailcone for transport to Toulouse. Airbus Industrie (consortium of four countries) were the benefactors of the collaborative experience gained from the Concorde project.

from British and French firms – BAC and Sud Aviation for the airframe, Bristol Siddeley and SNECMA for the engine – which were to make detailed proposals for carrying out the programme. There was no break clause. The agreement was signed by Julian Amery and Peter Thomas (Parliamentary Under-Secretary of State, Foreign Office) for the United Kingdom and Geoffrey de Courcel, French Ambassador in Britain, for France.

It was the first time in history that such a collaborative organisation had been set up.

The bureaucratic machinery that was to control the project stipulated that every decision was minutely and critically examined by both partners. This resulted in the project taking twice as long, and costing twice as much as it might have done had one partner had design leadership. Nevertheless the outcome, Concorde, was an astonishing technical achievement. It had the admiration of the world, though as it turned out, it was not to have the world as its customer.

The first flight of the prototype was expected during the second half of 1966, while the first production aircraft was due at the end of 1968, with a Certificate of Airworthiness (C of A) being granted at the end of 1969. In the event, the first prototype flew in March 1969, and the C of A was granted in late 1975.

The development costs of the project in 1962 were put at between £150 million and £170 million. The division of responsibility was to be 60:40 between France and Britain on the airframe and 40:60 on the engine. For various reasons connected with inflation, devaluation of the pound, the reworking of the design in 1964 for just the long-haul version and the longer time it took to complete, the project eventually cost, as we shall see, rather more!

Britain was not yet a member of the Common Market. As one commentator put it 'Concorde was conceived before the two contracting partners were joined in wedlock.' In spite of the agreement, membership of the European Economic Community (EEC) by Britain was put off for a further nine years. So even this act of European unity, in the face of American aviation technology, was not enough; de Gaulle was still suspicious of Britain's special relationship with the United States.

TAKING SHAPE

An SST has to operate in an extraordinarily hostile environment. At Mach 2 it experiences a freezing airflow but of such force that the fuselage is heated to the boiling point of water and shock waves tear at every angle on its airframe – and all this occurs at an altitude where the atmospheric pressure is a tenth of its value at sea level. Only military aircraft flown by pilots, equipped with sophisticated oxygen supplies and wearing pressure suits, had ventured to these extremes and then usually only for minutes at a time. The SST was designed to carry ordinary airline passengers in complete comfort and safety for hours at a stretch. Could the fusion of British and French philosophies possibly produce such a craft?

Judging from the expected timescale of the project there was confidence that success would come quickly. The physics of obtaining lift throughout the speed range was understood. The loss in aerodynamic efficiency at speeds above Mach 1 would be more than compensated by the improvement in the efficiency of the intake and engine combination achievable at Mach 2. A certain amount of ingenuity would be called for in some areas – notably in the engine air intake, air conditioning and fuel systems. Novel techniques, like the redistribution of the fuel in flight to maintain the balance of the aircraft, would have to be developed.

The aircraft also would need to behave in a way that pilots had been trained to expect. A slender delta-winged aircraft with no horizontal tail surfaces is less 'speed stable' so behaves differently from a swept-winged aircraft with a tail. A conventional aircraft is designed to pitch up if the speed increases and vice versa to restore the original airspeed. This effect is achieved with no control inputs, through the tail surface being forced down aerodynamically due to the increasing air flow and vice versa. Also the delta's ratio of pitch to roll is different from a conventional aircraft because of its length and small wingspan. Rather than rely on excessive training time to teach pilots new techniques it would be safer and much smoother to fit the aircraft with computers to give it the feel of a conventional aircraft.

The design was revolutionary. In spite of the very best endeavours of the designers there was no way of predicting the interaction of all the systems before the first flight of the prototypes. There might be some totally unforeseen side effect which could take years to iron out, like some quirk in the airflow affecting just one engine, necessitating a control condition peculiar to that engine. Indeed such a problem did arise with Concorde, but mercifully a simple solution was found.

Making the structure safe, yet light, would stretch the powers of the designers. Having built the structure it would have to be tested to limits beyond those that could be explored in flight. To achieve that, two full scale versions of the airframe would have to be dedicated to ground testing. All the stresses and strains

Above: An Avro Vulcan with five Olympus engines. The centre one – the Olympus 593 – is shown here undergoing icing trials. Water droplets, simulating flight in cloud, were released ahead of the intake from the device beneath the nose. An unchecked build up of ice within the intake and over the first compression stages of the engine would radically affect performance. Heat applied either electrically or from hot air on vulnerable surfaces overcomes the problem.

of high-altitude flight would be applied to them, including the repeated heating and cooling of the airframe, experienced on each supersonic flight.

Very high altitude flight is advantageous to an SST since it can fly at a far lower indicated airspeed for a given true airspeed (see page 11). At 60,000ft travelling at Mach 2, or 1,150 knots true airspeed, Concorde only 'feels' 435 knots of airspeed. Fighter aircraft have to be

Opposite: A production Concorde in flight, illustrating the delta's lack of horizontal control surfaces at the tail. Note the lengthened tail cone in comparison to the prototypes. (© Alan Schein Photography/Corbis)

built with sufficient strength to withstand turbulent air at indicated airspeeds in excess of 700 knots. Such a speed, apart from being quite unnecessary for an SST, would call for a stronger and therefore heavier structure. With the very small margin available of payload weight to total weight, the SST could not be allowed to have the luxury of a high indicated airspeed limit. The indicated airspeed limit on Concorde varies with altitude, reaching a maximum of 530 knots at 43,000ft. A typical limit for a subsonic passenger jet is 350 knots.

The disadvantage of flying at very high altitudes (compared to a subsonic airliner) is that the fuselage has to be made strong enough to withstand high differential pressures. To give a cabin altitude of 6,000ft when the aircraft is at 60,000ft requires a differential pressure approaching 11lb per square inch (psi). With an 8psi limit, typical for most subsonic airliners, the cabin 'altitude' would be over 7,000ft when flying at 37,000ft.

Above: The full-scale wooden mock-up of Concorde at Filton, England. This was used to evaluate emergency evacuation systems as well as being an example of the appearance of the finished aircraft. In 1976, when the first BA crews were being trained, this hangar also housed a plywood Concorde flight deck complete with droop nose – a British design responsibility. Fully equipped with instruments and electrics it was originally used for evolving an ergonomic layout. Largely responsible for the transformation of the prototype flight deck into one suitable for ordinary airline use was BOAC's Flight Development Manager Captain Jimmy Andrews. At first and unprecedented at that time, the project involved the airlines interested in acquiring Concorde and various pilots' unions. This airline participation diminished when their options to purchase evaporated. Ultimately only the teams from Air France and BA remained.

Large aircraft require a multitude of systems: electrics, hydraulics, navigation, undercarriage and flying controls to name but a few. Systems fundamental to the safety of the aircraft must have back up systems. Some systems on Concorde were to have additional back up systems. For instance, Concorde was fitted with four methods of lowering the undercarriage: two of them from different hydraulic sources; one by letting it down by its own weight; and a fourth by use of compressed air.

Before passengers could be carried every conceivable combination of failures and consequent recourse to back up systems had to be explored. Might the warning of a failure of one system cause the pilots to take

Opposite: Concorde 002 under construction, Filton 1968. Note the deployed speed brake at the tail cone. This was fitted for deceleration from Mach 2 without thrust reduction, thus avoiding engine surges. At a Paris Air Display 002 performed with the brake extended; it was said so that it could be differentiated from 001, airborne simultaneously. In fact it was left out in error. For deceleration on the ground 002 was fitted with a braking parachute. (Airbus)

Concorde 002 interior, *c.* 1970. The Flight Test Observers are wearing immersion suits, pressure suits and pressure helmets. The chances of bailing out at Mach 2+ from 60,000ft and surviving the experience, even if equipped for a space walk, would have been remote. The prototypes and pre-production aircraft were fitted with downward facing escape hatches held closed with explosive bolts. Should a hatch be 'blown' at 60,000ft, air would have been sucked out and the cabin would have 'climbed' to 100,000ft. On later test flights the crews abandoned such survival clothing. Once during pre-flight checks an escape hatch was inadvertently blown, luckily no one was hurt. (Airbus)

inappropriate action? Just prior to Concorde entering service it was discovered that the loss of some of the hydraulic pressure to the flying controls during the early stages of the take-off run resulted in an instruction to the pilots to abandon the take-off. This required the application of the wheel brakes, themselves requiring hydraulic pressure supplied from the source whose failure caused the take-off to be abandoned. Although hydraulic pressure to the brakes could be very quickly

Concorde 001 in its joint Aerospatiale/BAC livery; note the shorter tail cone and the different secondary nozzles in comparison to the production aircraft.

restored, it would not have been quick enough to prevent Concorde at 160 knots (270ft per second) from running off the end of the runway.

Safety at the level of system failures was one thing, but would the SST be able to survive a potentially more catastrophic incident? Would it be controllable at Mach 2 following the simultaneous failure of two engines on one side? The answer had to be 'yes'. But when the British and French prepared to join forces in 1961, experience of such incidents had not been good.

The world's first big supersonic jet was the four-engined, delta-winged American Convair B-58 Hustler. It was not capable of surviving a single engine failure at Mach 2, let alone two on the same side. Even flight at other speeds was risky – a B-58 crashed during a display at the 1961 Paris Air Show. Deaths of B-58 crew were common occurrences in spite of their being housed in escape capsules. It would be small comfort for a Concorde passenger to think that the pilot and his crew were similarly protected while he was not.

No country had yet been called upon to award a C of A for the transport of the fare-paying public in an SST. New rules had to be introduced. Every detail of the new aircraft would suffer the expert scrutiny of the civil aviation authorities of Britain, France and, as a prospective purchaser, the United States.

There are conflicting requirements on any aircraft. On an SST such conflict is extreme: too strong and it would be too heavy; too efficient at one speed, it might be unflyable at another; highly efficient engines at Mach 2 might be noisy on take-off. Compromise and ingenuity would find solutions, but the margins were narrow. The payload to maximum take-off weight on the long-range airliners of the 1960s was 10%; the Boeing 747s of the 1970s gave 20%. The SST would be lucky to achieve 5% over a 3,500nm (nautical mile) range. A tiny error in performance would remove the ability to carry any payload between London and New York.

The small predicted payload to weight ratio did not daunt the supporters of the SST. The difference would be restored by the greater number of Atlantic crossings achievable by one aircraft. The price of fuel was

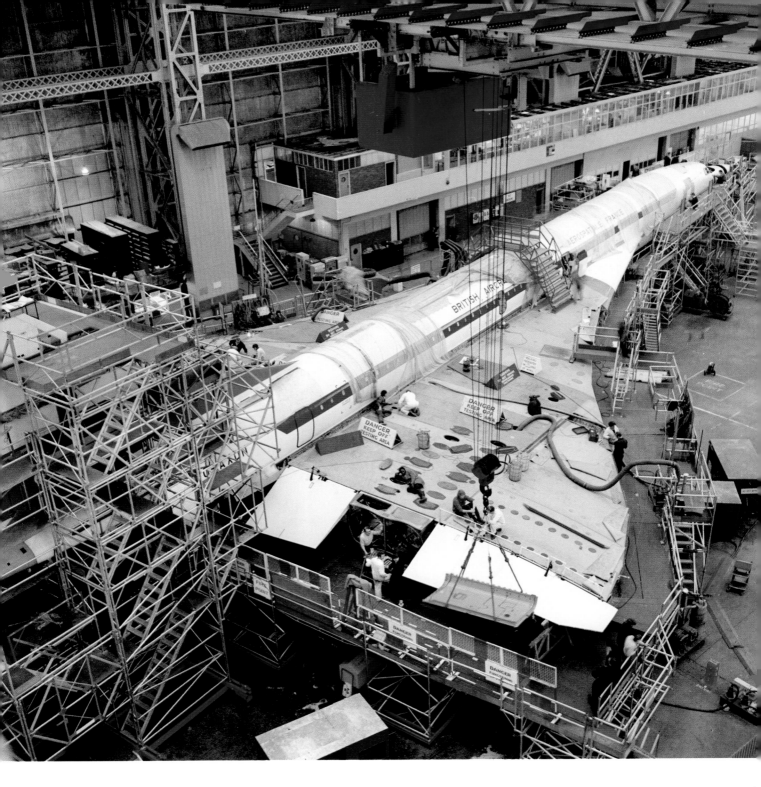

Concorde 01 – the first pre-production aircraft inside the 'Brabazon' hangar at Filton, Bristol.

expected to remain constant, even fall a little! The sonic boom, however, did cause concern.

An aircraft travelling at speeds above Mach 1.15, depending on atmospheric conditions, causes a double boom to be audible on the ground. Should this preclude supersonic flight over populated land, the medium–range

SST espoused by the French would be a non-seller, as there are too few viable medium-range routes which fly over the sea.

In 1966 the designers hoped that two very similar SSTs would be built. The medium-range version would have space for more passengers and less fuel whilst the long-range version would reverse the mix. It turned out however, that each design compromised the other. Coupled with thoughts about the sonic boom, this finally killed off the medium-range SST.

Nevertheless the two prototype Concordes 001 and 002, whose building had commenced in April 1965, could not be described as long-range SSTs. The same applied to the two pre-production Concordes – 01 and 02. The final production version of Concorde (serial numbers from 201 to 216) was not only a long range SST but also came very close to Archibald Russell's original specification – the Bristol Type 198. Designing, tooling and building three different versions of Concorde added very greatly to the cost and the time spent on the project. Additionally, mainly for political reasons, the whole project had duplicate headquarters, at Filton and Toulouse. With hindsight it is absolutely remarkable that such a fine aircraft as Concorde could have had such a complicated beginning.

No less troublesome to the project were the various changes of government on the British side. No sooner as one government takes over, its predecessor's activities

The first Concorde ordered by BOAC was given the registration G–BOAC. By the time it was delivered in 1976, the airline had become BA. BAC Weybridge built the forward fuselages.

come up for scrutiny. The Labour Government which took over from the Conservatives in 1964 was a case in point. It was only the agreement between France and Britain jointly to build an SST – an agreement that had been registered at The Hague – which prevented cancellation of the project.

Pressure had also been mounting from America for Britain to abandon various aviation projects – among

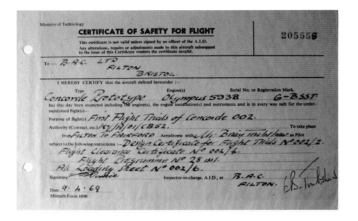

Left: This certificate of safety for flight allowed Brian Trubshaw to fly Concorde 002 from Filton to Fairford at the start of 5,500 hours of the flight test programme (the most tested airliner ever) shared between the two prototypes, two pre-production and the first two 'production' aircraft.

Below: A full-sized rig was made to check the working of Concorde's fuel system. Apart from supplying the engines, the fuel is used to move the centre of gravity rearwards during acceleration and forwards prior to landing as well as being a medium for cooling the air in the air conditioning system and cooling the engine oil.

The press statement for this photograph read: 'Concorde 01 on the move... the first pre-production aircraft was today (31 March 1971) moved from the Aircraft Assembly Hall at Filton to the weighbridge for weighing, as a preliminary to a month long programme of resonance tests.'

them the TSR2 (Tactical Strike Reconnaissance) supersonic military aircraft. Why should the United States lend money to Britain only to fund aircraft projects which would compete with the American aviation industry? According to Julian Amery, the British minister who had concluded the agreement with France, the Americans had insisted that Concorde should be stopped. George Brown, the Minister of Economic Affairs of the new 1964 Labour Government denied that this was so. Following publication of Geoffrey Knight's book *Concorde: the Inside Story* in 1976, Lord George Brown accused Mr Knight (in a

letter to *The Times*, 7 May 1976) of getting 'almost every strand of his story wrong', although he admitted that he had not read the book. Nevertheless Geoffrey Knight later confirmed that there had indeed been US pressure against Concorde.

At industry level the story was different. From Boeing (the company chosen to build the US SST) came support for the European project. At first sight it might seem odd that a rival should support a rival, but this support was mutual. Each company felt that if its rival was allowed to continue, then its own chances were bettered. This was borne out by the plethora of difficulties that faced Concorde as soon as the United States SST project was cancelled in 1971. Sadly much of that difficulty came from ill-informed political opportunists. Happily that lobby became far better informed.

On 2 March 1969, the day before I joined BOAC, these difficulties lay in the future. On that day, André Turcat (director of Flight Test Aerospatiale) was to attempt to take Concorde 001 into the air for the first time from Toulouse. As an observer of the television coverage I can well remember the occasion. Bad weather had delayed the flight, but at last Concorde 001 was lining up for take-off. About 30 seconds worth of take-off run would prove whether the preceding seven years had been spent in vain; they were not, she flew. It was an amazing moment.

Left: André Turcat in Concorde 01 F-WTSS, which became, under his command on 2 March 1969 from Toulouse, the first Concorde to become airborne. Compare the prototype with the 'production' version of the flight deck on page 104–105. The four 'fire handles' are here positioned on the central portion of the 'glare shield'; later versions have automatic control switches, the fire handles being moved to the overhead panel within reach of the Flight Engineer.

Below: On the flight deck of Concorde 002 Brian Trubshaw (Chief Test Pilot BAC) and John Cochrane (BAC Test Pilot) – the pilots of the first test flight of 002, 9 April 1969. Note the large empty box at the centre of the instrument panel – space for a 'moving map display' which was later removed. In 1974 John Cochrane was responsible for landing 002 safely after an over-quick extension of the left main gear broke its main side stay. The gear was strengthened after this incident.

Only the low speed and low altitude range of the flight envelope (combination of speed and altitude at which an aircraft is allowed to fly) were to be explored on this day. Did the indications of speed and altitude correspond to those of the chase aircraft? The undercarriage was not even raised. Take-off and flight proved to be possible, but could the slender delta be made to land?

Some 35 minutes later Concorde 001 was on final approach at Toulouse, and all looked well. On touchdown, two clouds of dust and smoke were whisked from the wheels into the vortices over each wing. Maskell

and Küchemann had been right: a separated airflow had given lift even to an aircraft as big as Concorde. Though fully aware that far better pictures would be immediately available I nevertheless photographed the event from the

Below: The complexity of the wiring is a wonder of the aeronautical world, let alone that it had to be designed before the aircraft was put together. It was said that the engineering content of the forward fuselage of Concorde was equivalent to an entire VC10.

Opposite: Concorde makes its airshow debut at the preview of the Farnborough Air Show, 7 September 1970. The auxiliary intakes on Concorde 002, just visible in this picture to the rear of the main intakes are designed to 'scoop' air into the engine. On the 'production' Concorde the scoop was replaced by 'a blow-in door' and the grey exhaust smoke trail was eliminated by redesigning the combustion chambers on the Olympus engine. (© Hulton-Deutsch Collection/Corbis)

Fairford, 12 November 1970, where 002 has just returned from having achieved Mach 2 for the first time. With Brian Trubshaw (orange flying suit) are most of the 300 personnel associated with the UK Concorde testing (engines, fuel, systems and field performance). Behind him is Bob Griffin (Snr Ops Officer, in a blue anorak). On Trubshaw's right are John Cochrane (Test Pilot), George Wood (Navigator), Brian Watts, Alan Hayward and Peter Holding (Flight Engineers). On Trubshaw's left, Roy Lockert (Nav), Eddie McNamara (TP), Doug Hawley (FE), Peter Baker (TP) – briefcase under arm and fresh from flying the Canberra 'chase plane', John Allan, Mike Addley, Dick Storey, Roger White-Smith – raincoat, and Mike Crisp (Head of Flight Test). Those in white overalls are the inspectors, in blue, the fitters. (Airbus)

television screen to have, as it were, a personal record of Concorde's first landing.

Some minutes later surrounded by cameras and microphones André Turcat announced in French and English, 'The big bird flies.' I hoped from that moment that one day I would have the privilege of piloting Concorde. It had been so exciting to see the take-off of the prototype of such an interesting project, that I longed to learn how it was done.

Just over one month later, the British-assembled prototype Concorde 002 took off from Filton with Brian Trubshaw (Chief Test Pilot of BAC) at the controls. John Cochrane was the co-pilot and Brian Watts the Flight Engineer. As the flight deck is over 35ft above the runway on main wheel touchdown, of key assistance in landing is the radio altimeter. Concorde has two radio altimeters, on this occasion both failed, so the landing at Fairford, Concorde's British test flight base (on account of its long runway), had to be done by eye. It was a 'firm' landing. As Brian Trubshaw put it later: 'We arrived about half a second early.'

Just under six years of testing and building were to elapse before Concorde's first commercial flight. During that time crises of all sorts presented themselves: environmental, political, technical and economic. Concorde's surviving of them all ranks as a modern wonder of the world.

THE OTHER SUPERSONIC TRANSPORTS

The flight deck of the Boeing 747 is situated above the first class passenger cabin. The decision to put it there was so that it would be easy to convert the airliner into a freighter. This could be done by fitting a cargo door in the nose and removing the seats. Boeing believed, during the design phase of the 747 in the mid-1960s, that supersonic transports would render their jumbos obsolete as passenger carriers, but not as freighters. They had good reason for their belief: the Concorde project had started in November 1962 and in June 1963 President Kennedy had announced that the US would develop an SST.

That announcement came after much lobbying of the President by the FAA (Federal Aviation Authority) administrator, Najeeb Halaby. He warned of dire consequences if the US was unable to build a challenger to Concorde, adding that the President could conceivably find himself flying in a foreign aircraft. In the same way that Russian success had spurred on the US space programme, Concorde, especially after Pan American announced its intention to buy six, catalysed the US reaction to SSTs. They would build one too.

To compensate for a late start the US SST had to be much larger and faster than Concorde. To this end a competition was planned, administered by the FAA, to look for the best airframe and the best engine. On 31 December 1966 it was decided that a Boeing design with swing wings (designated B2707-200) and powered by four General Electric engines was the aircraft to carry the US into the supersonic passenger age. The runner up in the competition had been the non-variable geometry Lockheed L-2000 – a proposal of similar shape to Concorde. The Lockheed design, being much simpler than Boeing's, would have appeared sooner which might have prevented the anti-SST movement from taking root. In April 1967 President Johnson gave the go-ahead for the next phase, a four year $1,600 million prototype programme. Unlike previous US civil projects this one was to be financed on a 90:10 Government to industry ratio, changing to 75:25 for cost over-runs.

Half as long again as Concorde, the B2707-200 at 318ft would have been the longest aircraft ever built. The design was intended to carry over 300 passengers at Mach 2.7 (about 1,800mph) over 3,900 statute miles at altitudes up to 70,000ft. Not only would its swing wings (fitted with conventional flaps and slats) give it a good take-off and landing performance, but they would give a higher aerodynamic efficiency at subsonic cruising speeds compared to a fixed delta design. Its maximum weight was to have been 675,000lb (306,000kg) almost one and three quarter times that of Concorde. Each of the GE4 turbojet engines would have been capable of producing nearly 70,000lb of thrust – not quite double that of Concorde. Due to the kinetic heating experienced at Mach 2.7, a titanium alloy would be required, such high temperatures (as hot as 260°C/500°F) being too great for aluminium alloys (see page 11).

By any definition it was an ambitious project, and it very quickly ran into difficulties. The hinge mechanism for the swing wings presented the greatest problem. For maximum effectiveness swing wings must have their pivots as close to the centre line of the aircraft as possible, since the greatest benefit of increased wing span can be thus achieved. However, this interfered with the undercarriage and the positions of the engines. For it to be worth having swing wings their associated machinery must not be too heavy. By 1969 it appeared that their weight might be such that no payload could be carried.

So Boeing, unable to continue with the swing-wing project, submitted another design for FAA approval. The B2707-300 was somewhat similar to Concorde but it had a more marked double delta wing and a tail plane. The same engines and as many of the original systems as possible were to be used. This time about 250 passengers were to be accommodated and the expected range was to be about 4,200 statute miles. But with a Mach 2.7 cruising speed, a titanium alloy, as yet not finally developed, would have been necessary. In comparison to Concorde's very simple arrangement of six elevon surfaces (the flying control surfaces), the B2707-300 had a tail plane as well as control surfaces on the trailing edge of the wing. As with Concorde, the design of the B2707-300 became more difficult as it progressed. By 1970 political and environmental opinion was hardening against SSTs. Their future looked uncertain.

A wind tunnel model of a supersonic transport. The 30ft by 60ft tunnel at the NASA Langley Research Center, Hampton, Virginia, USA, was used in the mid-1970s to test the model at low speeds. (© Corbis)

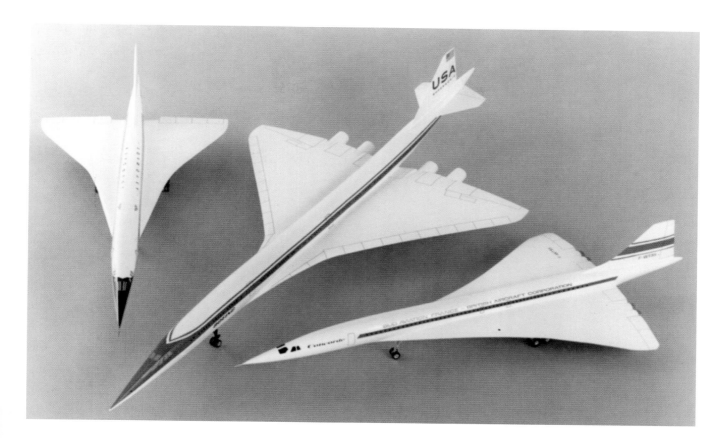

Of the three SSTs only Concorde became successful. The swing wing B2707–200 having proved too ambitious a project, Boeing resorted, in 1969, to the 2707–300 version (shown here as a model at centre). By 1971 the US government withdrew support even for this programme. On the left is a model of the first version of the Soviet Tu-144 and on the right a model of a prototype Concorde showing the original visor – fitted only to the two prototypes.

The environmentalists found allies among the US politicians critical of government expenditure on civil programmes. Had the same politicians withdrawn financial support from environmental programmes, then there would have been no joining of forces. As it was, both parties wanted the US SST to be cancelled. Notable among the politicians was Senator William Proxmire. Among the environmentalists were groups that delighted in such names as 'The Committee for Green Foothills' and 'Friends of the Wilderness'.

On 24 May 1971 they achieved their victory. The Senate and the House of Representatives both voted against further funds being made available for the US SST. For a sum of money about equal to that spent by Britain on Concorde, the US aerospace industry had nothing to show but tons of paperwork. Boeing's workforce at Seattle was drastically cut. Bleak though it was for Boeing, the Concorde protagonists knew that their task was now harder. Without an American contender the full force of the environmentalists, flushed with victory, would be focused on them. Even the presence of the Soviet SST (the Tu-144) would not count much in Concorde's favour. The technological mood in America, galvanised into such an intensive exploration of space by early Russian success had been assuaged by the moon landings. The Tu-144 now appeared a relatively minor threat to prestige. The new creed among some in the US seemed to be: 'If you cannot beat them, then question the environmental impact and economics of their projects.'

In Russia nobody appeared to question the irony of an egalitarian state producing an aircraft suitable only for commissars. The Tu-144 was not a copy of Concorde although superficially it looked very similar.

It was supposed to be able to carry 121 passengers at Mach 2.35 (1,550mph) over a distance of 4,000 statute miles – a performance which, on paper, was slightly superior to that of Concorde. 'Concordski', as the Tu-144 was dubbed, had a less complicated wing than her European rival. Instead of the ogival form with the curved leading edge joining the two angles of sweep-back, as on Concorde, the Tu-144 was more

Above: The Tu-144LL deploys its drag 'chutes on landing. (NASA/IBP)

Below: The Boeing 2707-200 of 1966. Note the presence of flaps and slats further complicating the engineering problems already associated with the swing-wings.

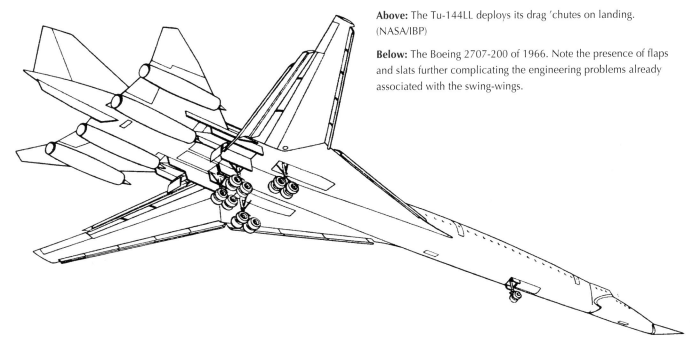

distinctly a 'double delta'. The wing also exhibited less camber droop and twist than Concorde's. The four long engine intakes for the NK 144 turbofan engines were arranged in one box underneath the centre line but with dividing walls between them. The main gear was away from this box, retracting into a bay which protruded into blisters above and below the wing surface. There were no underfloor holds; baggage and freight were to be carried in panniers stowed between the two cabins and in a compartment to the rear over the engines.

The Tu-144 however chalked up several 'firsts'. On 3 December 1968 she made her maiden flight, two months ahead of Concorde. In June 1969 Mach 1 was exceeded and in May 1970 Mach 2 was achieved. The respective dates for Concorde 001 (F-WTSS) were October 1969 and November 1970. However, there

were problems with the Russian design as Sir George Edwards, Chairman of BAC, had pointed out to the Russians in 1967. The engines were in the wrong place, too close inboard; presumably they had been so positioned to make the aircraft easier to control in the event of engine failure. The engines were 'turbofans' and this would impair efficiency during supersonic cruise. The wing was not sufficiently sophisticated which would impair effectiveness and efficiency throughout the speed range.

At its appearance in the 1973 Paris Air Show, the Tu-144 looked to have been radically redesigned. The pairs of engine intakes were now in two separate boxes, placing the engines further outboard, while the main undercarriage legs now retracted into a compartment within the engine intakes. A retractable 'canard' or foreplane appeared, placed above and just to the rear of the flight deck. Extended, the canard would improve low speed flying characteristics by giving a lifting force to the front of the aircraft. This would now be countered

A model of the Tu-144 in a giant wind tunnel. (RIA Novosti/aviation-images.com)

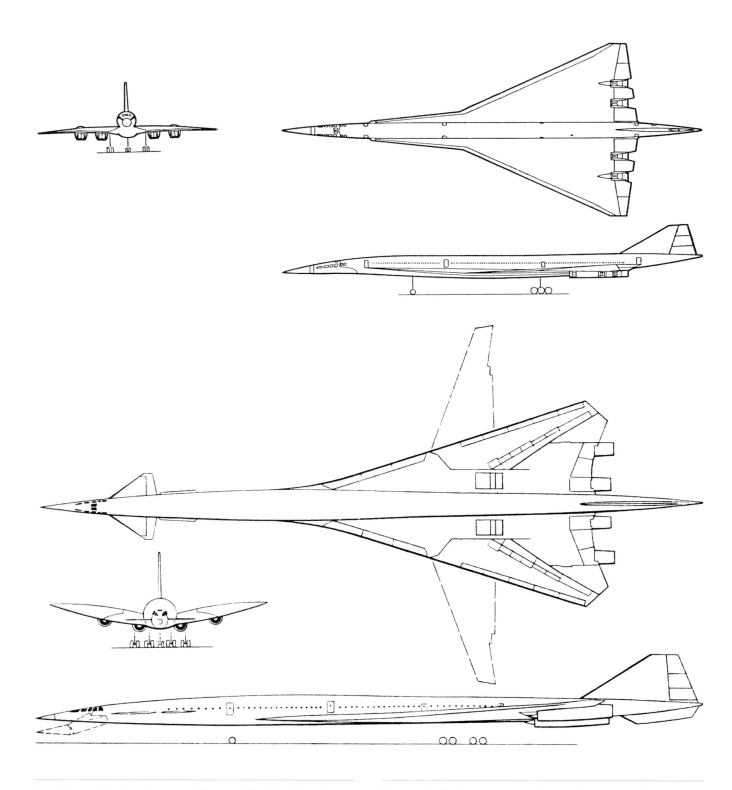

Top: History might have been very different had Lockheed's much simpler proposal for an SST, the L2000, been chosen instead of Boeing's highly complex swing-wing version. Since 1960 Lockheed had gained immense supersonic experience from the SR-71 'Blackbird' (Mach 3+) project. So they might have had a prototype L2000 built before the mood against a US SST had gained momentum.

Bottom: The Boeing 2707-200. By the time this version of Boeing's variable-swept-winged SST had appeared, grave doubts as to its viability were being voiced. Here a canard has appeared to overcome shortcomings in pitch control.

by the elevons to the rear of the wing controlling the aircraft in pitch and roll, going a few degrees down to give 'flap' effect to the wing. Retracted, the canard would not interfere with the supersonic airflow. The Boeing 2707-200 (swing-wing) design eventually included a canard too; but it was non-retractable and was there to assist control in pitch. To be fair, Concorde also appeared in progressively superior forms, but never underwent such a radical change as that exhibited by the Tu-144.

Tragically the new version of the Tu-144 crashed at the 1973 Paris Air Show. It appeared to be attempting to recover from a dive. One theory suggested that the engines had flamed out following a harsh manoeuvre which had been initiated to avoid a French Air Force Mirage. The pilot then dived the airliner to 'windmill' the engines to a sufficient rotational speed for a successful relight (restart). On seeing how low he was, he pulled back and overstressed one wing which broke off causing the aircraft to invert and break up. On 26 December 1975 the Tu-144, which apparently had not been modified greatly following the tragedy, entered service between Moscow and Alma-Ata, capital of Kazakhstan. It carried mail and freight over this 1,800 statute mile sector and flew at about Mach 2.05, between 52,500 and 59,000ft for much of the distance.

Left: A good view of the underside of the later version of the Tu-144. Compare the arrangement of the undercarriage, the intake boxes and the presence of a canard with the earlier version of Tu-144.

Below: The Tu-144 above the clouds. (RIA Novosti/aviation-images.com)

Above: A Tu-144D (background) is framed by the distinctive drooping nose of another Tu-144, the LL 'Flying Laboratory'. (NASA/Jim Ross)

Below: The first version of the Tu-144 had the four separate engine intakes ducts contained within a single box. Note the absence of a canard. Although superficially similar to Concorde's wing in two dimensions, the Tu-144's wing lacks the complex three dimensioned shaping found on Concorde.

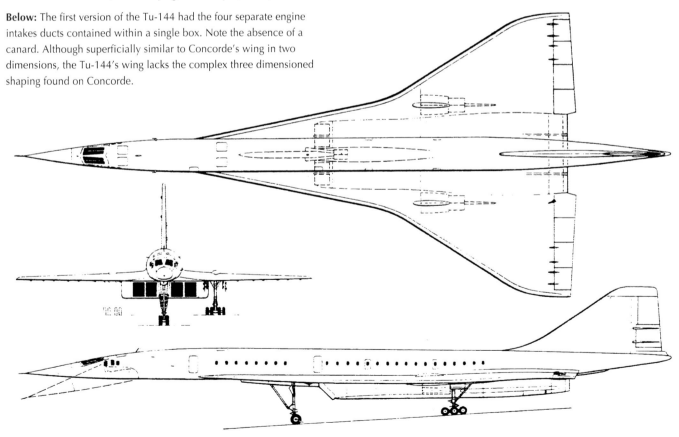

Above: On 29 November 1996 the Tu-144LL (RA-77114) flew again from the Zhukovsky flight test centre, near Moscow. A series of 32 joint Russian–US test flights in the High Speed Research study program were originally proposed. Out of storage after ten years, this aircraft, one of the 17 originally built, was re-engined with Kuznetsov NK-321, two-shaft turbofans, replacing the Koliesov design bureau engines, which were fitted to only five of the Tu-144s for longer range. (NASA/IPB)

By 1985, if not years before, the Tu-144 was out of service. There were unconfirmed reports of another accident but such events, unless they involved foreigners, were usually cloaked in Soviet secrecy. There is no evidence to believe that the Tu-144 ever achieved its design range and payload. On the contrary, it almost certainly had to use the afterburners (reheat) during its cruise. The low compression ratio associated with an afterburner makes it inefficient. At the cost of doubling the fuel flow on Concorde, an afterburner gives only 20% extra thrust. Without an afterburner Concorde would have needed heavier engines to give the thrust required for take-off and acceleration through Mach 1. The Tu-144 not having such a sophisticated variable area intake may not have had

been able to switch off its afterburner until it became light enough (having used up fuel) at the latter end of the supersonic cruise.

The Tu-144 did not have sufficient range to cross the Atlantic. So the reciprocal plans for the Tu-144 to fly from Moscow to New York via London and Concorde to Tokyo via Moscow and Novosibirsk had to be shelved. The inability of the Soviet system to equal or be superior to the West's engineering skills was not to be admitted or drawn attention to. Nevertheless data for the airfield at Novosibirsk remained on the Concorde simulator at Filton, acting as an inauspicious reminder of what could have been.

Overleaf: 'The woods decay, the woods decay and fall, The vapours weep their burthen to the ground, Man comes and tills the field and lies beneath, And after many a summer dies the swan.' Lines from Tennyson's *Tithonus* and applicable to the Soviet technology decaying in a Moscow wood, as seen by Mark Wagner (photographer for *Flight International*), September 1993. In March 1996, after ten years of storage, a modified Tu-144L with new Kuznetsov NK-321 turbofans was rolled out at Zhukovsky air base for six months of joint Russian–US (NASA) flight research. (Mark Wagner/aviation-images.com)

A STORMY BEGINNING

Early in the 20th century, self-propelled road-going vehicles in England had to be preceded by a man with a red flag. The rule did not reflect the politics of the day but was made because of the vehicles' speed. Whether those who criticised Concorde were motivated by this type of conservatism will never be known; nevertheless there was vociferous opposition against SST's in general and Concorde in particular. The criticism came from all quarters: journalists (on both sides of the Atlantic), an English bishop, politicians (mainly from America), and many other individuals. Their arguments were based on economic and environmental grounds and were often completed with the question: 'What's the use of people travelling so fast in any case?'

When the American SST was cancelled in 1971, a US Senator declared that if the project was worth financing then Wall Street, not the Government, should do it. By 1976 Concorde had absorbed in development costs alone about £500 million from Britain and the same amount from France – rather less it was said, than that spent on the abandoned US SST. Concorde certainly had rather an expensive aura to it. Nevertheless it would be churlish for any American politician to maintain that the successful US civil transports, like the series of Boeing and Douglas passenger jets, had not benefited financially from the military contracts that paved the way for their development. But with no military counterpart, the US SST as well as Concorde had to be developed and paid for almost from scratch.

Concorde's development bill had suffered in three ways: firstly from inflation; secondly from the costs associated with having two equal partners (according to one source this accounted for as much as 30% of the bill); and thirdly from having produced three substantially different versions of the aircraft – the two prototypes, the two pre-production aircraft and finally the production series. It was easy to see the critics' point of view: that vast amounts of money had been spent so that the 'idle' rich could save a few hours travelling time. But Concorde was never designed for

Michael Heseltine MP addressing Sir George Edwards (left), Brian Trubshaw (in the flying suit) et al. Judging by their facial expressions, the 1972 World Tour with Concorde 002 has not produced the orders which were hoped for.

Above: Concorde 01 with shockwaves visible in the supersonic exhaust flame. The prototypes had the 'petal' type secondary variable nozzle. For reverse thrust separate buckets angled the exhaust flow forward through louvres positioned top and bottom of the jet pipe. The later version of secondary nozzle combined the three functions of silencer, divergent duct and thrust reverser. The British Concorde was test-flown from Fairford. However, the New York noise abatement procedures were evolved in Casablanca (Morocco, Latitude 34N). There, proximity to the equator was also useful. With the higher tropopause over the equator than over the poles (c. 55,000ft cf. 30,000ft), the temperature drops as low as -80°C (cf. -45°C). Concorde, suddenly encountering cold air during a demonstration to the Shah of Iran, shot into a barely controllable climb. Curing this alarming characteristic nearly delayed Concorde's entry into service.

the idle. On the contrary it was designed for busy people who added to the wealth of their businesses and the economies of the free world. Given that each generation benefits from the preceding one, would every Concorde critic, had he been in a position to do so, have protested at the development of the car in the days when it was exclusively used by a minute proportion of the population?

Even as late as 1976, critics were pressing for Concorde to be abandoned on economic grounds. But by then the development phase was virtually over, so cancellation would have ensured that all the money spent and experience acquired would have been largely in vain. Those who suggested that Concorde's development money should really have been allocated to other more 'worthy' causes were being over optimistic. It is unlikely that the small amount that would have been made available by cancellation in

Below: A pollution chart published in 1976 showing the vey small amounts of pollution attributable to jet engines, let alone SSTs.

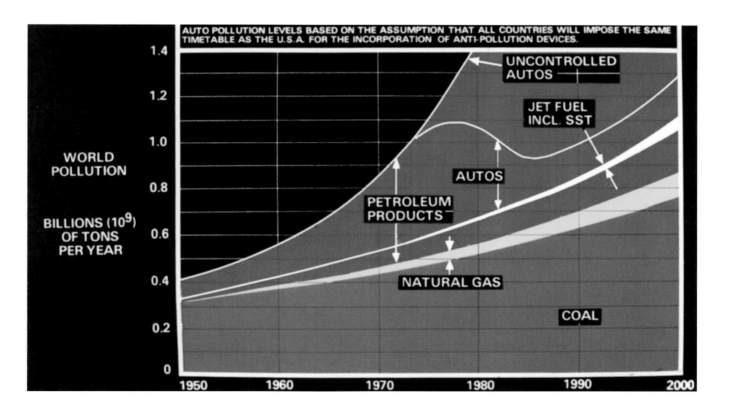

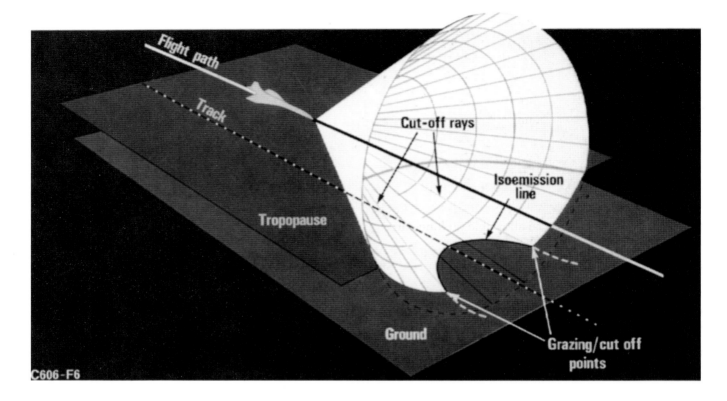

The direction of the 'ray' path of the shock wave is at right angles to the line of the wave. As the shock waves descend into the warmer air near the earth's surface their path bends concave to the sky. Thus shock waves generated by aircraft at low supersonic Mach numbers are not heard on the surface, having been refracted upwards, in a way analogous to a mirage.

1963 (about £40 million) would, as a matter of policy, have been channelled elsewhere.

A more serious criticism is that Concorde prevented investment in other aviation projects. Would the projected BAC 3-11 (a wide-bodied, twin-engined, medium-range airliner) have survived in the early 1970s? Or would Britain have remained a full member of the European Airbus consortium? As for the BAC 3-11, the answer is probably 'yes' bearing in mind that governments often withdraw their support from aviation projects; but without the agreement to build Concorde, the Airbus consortium would not have come into being when it did, so the second question is hypothetical.

Alongside their economic arguments, critics added environmental ones. One criticism was that the ozone layer might be dispersed and the ultraviolet light shield removed thus causing skin cancer to become endemic. Others mentioned physical damage caused by sonic booms, passengers suffering excess doses of radiation from solar flares, pollution of the atmosphere with poisonous emission and finally Concorde being too noisy on or near the ground.

Concorde's manufacturers patiently answered all these questions. On the first point, ozone is present in the stratosphere, which is the layer of atmosphere above the tropopause. Between the tropopause (average altitude 37,000ft) and the surface of the earth, is the troposphere, which is that part of the atmosphere where weather occurs (see Appendices). Concorde's engines emit nitric oxide which reacts with ozone to form nitrogen dioxide. Without cleansing from the troposphere, the stratosphere would lose ozone and gain nitrogen dioxide. It was feared that without absorption by the ozone the extra ultraviolet light from the sun could cause a higher incidence of skin cancer. However, an American investigation, called the Climatic Impact Assessment Program (CIAP), refuted the current theory that the 30 SSTs then scheduled to enter service would have any detectable effect on the stratosphere. Furthermore natural variations in the quantity of ozone would make the assessment of the impact of even 125 Concordes flying four hours per day above 50,000ft impossible to discern.

On the sonic boom problem, the British desire to build only a long-range SST was based on the belief that sonic booms would usually be intolerable over populated land, but would be allowable on long-range routes which were over oceans, desert or tundra. On some others a small detour adding maybe 100 to 200 miles (five to ten minutes flying time) could allow continuous supersonic operation through avoidance of 'boom sensitive' land. Where there was no way round, the higher subsonic cruising speed of an SST would give it a small edge, about 100mph, over the majority of subsonic aircraft. Most Concorde routes used to include a small proportion of flight at Mach 0.95.

On the question of solar flares, Concorde was fitted with a radiation meter. The usual dosage rate of cosmic radiation over the latitudes between London and New York at 55,000ft is around 1 millirem per hour (about twice that found at 35,000ft). Solar flares could cause the dosage rate to exceed 50 millirem per hour. In every case so far, the presence of radiation from a solar flare has been forecast since the flare can be seen about a day before the associated radiation reaches the earth. If radiation were encountered Concorde would descend to an altitude where, shielded by more atmosphere above, the radiation rate is lower. A survey used in 1976 showed that such evasive action would have had to have taken place five times during the previous 39 years. No radiation warnings from space were received on Concorde during the first ten years of commercial operation. Although the dosage rate on a subsonic aircraft is half as much, the occupants

A portion of the upgoing wave can be refracted back to the surface forming a secondary boom. During the First World War, the guns of the Western Front in France were heard faintly in certain places in England, on still evenings, further from them, than in others, closer to them, due to this effect. During the winter when the upper westerly winds are stronger, eastbound Concordes decelerate through Mach 1 further from a 'boom sensitive' coast to minimise this effect.

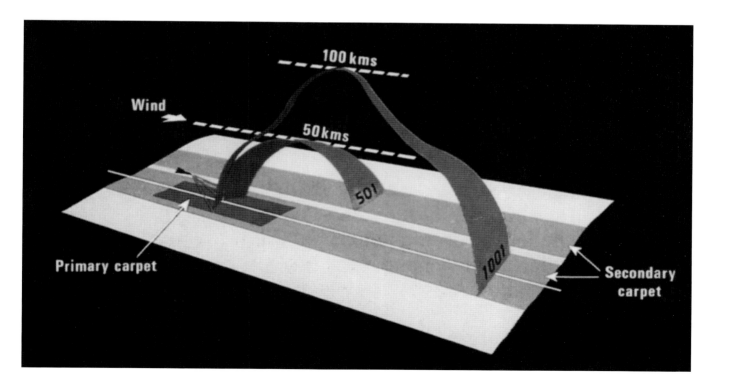

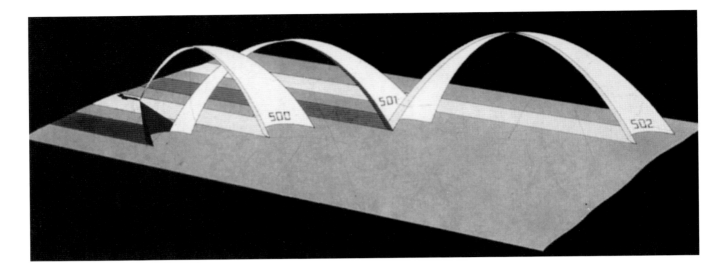

Under some conditions the shock wave can bounce, as shown here. The amount of energy returning to earth from these effects is tiny. Loose windows can sometimes be rattled by them, as by a gust of wind. The reflected and refracted boom is usually inaudible unless the ambient noise is virtually zero.

experience the effect for twice as long, thus receiving a similar quantity to that received by a Concorde passenger. Furthermore subsonic flights over the Poles, where the radiation is higher, possibly exposes the aircraft's passengers to more radiation.

Placing this all in context, an individual's dose of radiation from all sources, including cosmic rays, X-rays, television sets and atmospheric nuclear tests accumulates to about 100–150 millirem per year, and one transatlantic crossing in a subsonic or supersonic transport adds 3 millirem to that total. There are greater dangers to be found from crossing the road than from this. Suggestions that Concorde should only be staffed by stewardesses beyond child bearing age just turned out to be stories designed to attract popular press headlines.

As far as the pollution of the atmosphere is concerned only hydrogen-fuelled engines are completely 'clean' – that is if the water vapour they emit is ignored. Fossil-fuelled engines also emit water vapour. Often the critics could not make up their minds whether the presence of extra water vapour would 'cut off the sun's rays from the ground and bring on a new ice age', or 'give rise to a "hot house" effect and overheat the earth'. In practice the earth's weather systems are infinitely more responsible for the distribution of water vapour than aircraft can ever be. As for other pollutants, cars are nearly ten times worse per seat mile than Concorde, and there are many cars.

Of the most serious concern to the critics was the noise that they believed Concorde would cause at airports. For efficiency at supersonic speeds SSTs had to be fitted with engines which have a high velocity jet efflux. The shearing effect of this efflux with the static air to the rear of an aircraft causes the major part of the noise. Consistent with retaining an efficient engine, Concorde's manufacturers had gone to enormous lengths to reduce engine noise. They had concentrated on two areas. One was to break down the sharp boundary between the moving and static air and the other was to try and increase the quantity of air flowing through the engine so that it would give the same thrust with a lower exhaust velocity.

The first line of attack saw the development of a device called 'spades'. These were designed to break up the efflux by protruding rectangular metal plates into the jet pipe to splay out the jet efflux during take-off, to be retracted when engine silencing was not required. The 'spades' appeared quite promising in miniature, but failed to be sufficiently effective on the full-scale engine, and were abandoned. Another device, a pair of 'buckets', was more successful. Consisting of a pair of doors rather like eyelids, they are set to the rear of the jet pipe of each engine, being visible from the outside of the aircraft. During flight at low speed they mix the static and engine exhaust air thus breaking up the noisy boundary. On landing they are used as thrust reversers by closing off the jet pipe and deflecting the efflux forward; in supersonic flight they form a divergent exit nozzle (see photograph on page 35).

The second line of attack on the noise front involved computer control of the relative speeds of two engine shafts allowing the engine to 'change gear'. This happens after Concorde takes off and allows a greater mass of air into the engine, thus causing the velocity of the jet efflux to be reduced without the loss of thrust. Such a process can only be applied after Concorde has sufficient forward speed for the intakes to swallow enough air; insufficient air could lead to a surge (backfire).

Whatever the manufacturers achieved in the way of noise reduction, the crews, having reduced to minimum thrust consistent with climbing to comply with noise abatement procedures, had to eke out every last ounce of the aircraft's performance. However, such procedures had not been finally and accurately formulated for use during the route proving flights flown by Concorde during the summer of 1975. Concorde was thus rather noisier than she should have been, with the result that the American anti-Concorde lobby found some rather willing allies in Britain.

Concorde 002 on take-off. The 'production' Olympus engine has a modified combustion chamber which eliminates these smoky emissions. Also changed were the auxiliary inlet doors from the 'scoop' (shown here), to the 'blow-in' type. The variable 'ramps' are visible in the intake entrances. The failure and throwing forward of one of these from number 4 intake and its ingestion by the number 3 engine during a test flight at Mach 2 on 001, caused a double engine failure. The subsequent safe return to base proved the soundness of Concorde's basic design.

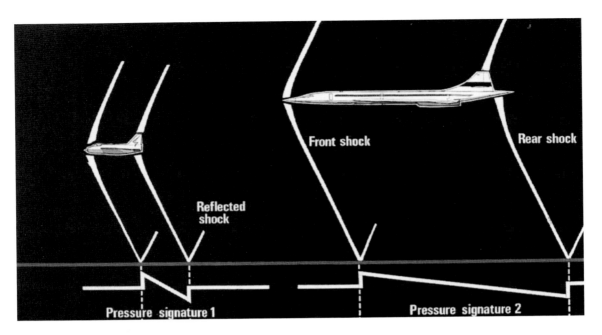

Front shock

Rear shock

Reflected
shock

Pressure signature 1

Pressure signature 2

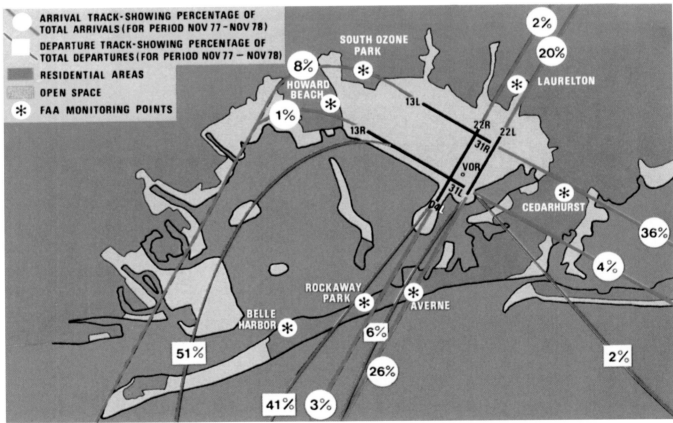

ARRIVAL TRACK-SHOWING PERCENTAGE OF
TOTAL ARRIVALS (FOR PERIOD NOV 77 – NOV 78)

DEPARTURE TRACK-SHOWING PERCENTAGE OF
TOTAL DEPARTURES (FOR PERIOD NOV 77 – NOV 78)

RESIDENTIAL AREAS

OPEN SPACE

✳ FAA MONITORING POINTS

Top: The pressure signatures from supersonic aircraft of different lengths. Note this diagram does not show the refraction of the shock waves due to increasing temperature and changing winds as they approach the earth's surface.

Bottom: The noise monitoring posts at Kennedy. The two most desirable runways for Concorde on take-off are 31L and 22R. By turning away from the noise sensitive areas after take-off, most of the noise impact of Concorde at Kennedy was avoided. Runways 4L, 4R and 13L were never used by Concorde at transatlantic take-off weights.

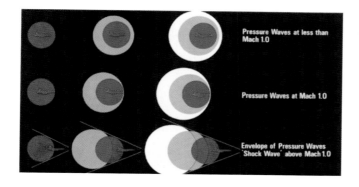

Above: The build up of the shock waves. As the aircraft approaches the speed of sound, the molecules of air progressively have less 'warning' of the approach of the aircraft. Above Mach 1 a shock wave forms, tangential to the spheres of disturbance. Some Concorde critics appeared to think that Concorde formed a shock wave at any speed. Under normal conditions, due to its refraction in the warmer atmosphere beneath the aircraft, the shock wave from Concorde only reaches the ground when Concorde exceeds Mach 1.15. The shock wave first reaches the ground about 50 miles after the 'acceleration point'.

Below: Still unwelcome at New York, Air France Concorde F-BVFB photographed on a stopover at a US Navy air station, 1977. (PH3 Caffro/US Navy)

The worst aspect of the anti-Concorde campaign was the way in which the critics were quite happy to see Concorde written off without being allowed into New York, and so without a fair noise trial. There was much discussion in the media about this but eventually, when services began, the combination of the relatively light take-off weights (less fuel is required on a New York to London sector than vice versa) and the meticulous application of the noise abatement procedures kept Concorde's noise well within limits. The protestors melted away. After that Concorde made many firm friends with the people of New York.

Nevertheless a second generation SST would have to be quieter than Concorde. Three or four fully laden Concorde departures per day were not only tolerated, but frequently admired. If more Concordes had been built major modifications would have been incorporated. Range and or payload would have been improved, but most significantly the next version would have been

quieter. The proposed 'B' wing Concorde might still not have been quiet enough to comply with the latest rules applicable to subsonic airliners, but would have been much more acceptable.

However, in January 1976 the final battle for entry to New York still lay in the future. British Airways (BA) and Air France (AF) were patiently putting the finishing touches to a plan for a joint take-off into the supersonic era; the British with a Concorde from Heathrow and the French with a Concorde from Charles de Gaulle.

On 21 January 1976 I stood with my wife, camera at the ready, at the western end of runway 28L at London's Heathrow airport.* Around us were crowds of people, some were equipped with headphones connected to noise monitors whose microphones were dangled from poles. Concorde would not start her commercial supersonic service unheard! We listened to a local radio broadcast. At 11.40am precisely the commentator announced that Concorde had begun to roll. Then we heard the distinctive roar of the four reheated Olympus engines. Finally Concorde G-BOAA came into view, landing gear retracting as she climbed to the west prior to turning to the south-east en route to Bahrain. Commercial supersonic services had begun.

---

* A note on runway designation: '27R' would be the rightmost of two west-facing runways. Runway '9L' at the same airport would be the leftmost of the two east-facing runways. Therefore '9L' and '27R' are the same runway, but referred to by different names depending on the direction in which it is being used. Runways are referenced to magnetic north, as are the centre line of airways and instructions from air traffic controllers.

As the British Concorde, under the command of Captain Norman Todd, with Captain Brian Calvert as second-in-command and Senior Engineer Officer John Lidiard, reached her subsonic cruising altitude over the English Channel, the crew heard that the Air France Concorde F-BVFA had had an equally successful departure. The odds, people had said, were heavily against achieving a simultaneous take-off, but as so often in the past Concorde had proved the pessimists wrong.

Concorde F-BVFA flew to Rio de Janeiro via Dakar; G-BOAA flew to Bahrain. Included amongst the guests on G-BOAA were Sir George Edwards (Chairman of BAC) and Sir Stanley Hooker. On arrival at Bahrain they were invited to a banquet at the New Palace by the Amir of Bahrain. Sir Stanley Hooker, in his book *Not Much of an Engineer* (Airlife, 1984), said: 'Eventually we entered the fabulous dining room, ritually washed our hands and took our places around the gigantic U-shaped table. I was next to the Minister of Foreign Trade, who spoke perfect English. We each had a waiter in full court uniform, each pair of guests being supervised by a steward... I was about to turn to the Minister and say "How far sighted and generous of your government to allow us to bring the Concorde here on its first scheduled flight" when he said to me "what a great honour you have done our country by bringing your magnificent Concorde here on its inaugural flight!" Considering the violent hullabaloo that was going on in New York, and that nobody else would allow us in, it was difficult to find the right answer!'

Oppposite: In contrast to her difficulties in 1976 and 1977 Concorde eventually became a welcome visitor to many US destinations. In November 1984 Concorde G-BOAB visited Boeing Field at Seattle, Washington State. Mount Rainier can be seen in the background. For the visit the Boeing Museum of Flight displayed a model of the cancelled US SST project, the B2707-200; they also organised a supersonic flight by Concorde out over the Pacific and back for 100 passengers to raise funds for the museum.

THE TURN OF THE TIDE

During the early days of supersonic services to Bahrain in 1976, Concorde was hardly ever out of the headlines. 'US warned against ban on Concorde'; 'Court challenge on Concorde go-ahead'; 'Red light delays Concorde'; 'Boeing tries for rival to Concorde.' Under this headline from *The Times* (3 April 1976) Mr Lloyd Goodmanson, Boeing's design director said '… much of the official opposition to giving British Airways and Air France landing rights for Concorde was based on pure jealousy of their commercial lead rather than environmental considerations.'

Concorde had many vociferous friends and enemies. But for Concorde to succeed she needed landing rights in the United States, and at New York in particular.

Two production lines for Concorde had been set up at vast expense: one in Toulouse, France, and the other at Filton, in Britain.

Apart from the availability and skills of a work force, production lines require the necessary jigs and tools, but above all factory space. In 1974 the Governments of Britain and France had approved the building of 16 production Concordes. This was in spite of Pan American and TWA having decided in 1973 not to take up their options on buying Concorde. It was hoped that now Concorde had started services, these airlines might change their minds.

Iran Air and China Airlines had also shown an interest in Concorde, but they, in line with the rest of the world's airlines, were keeping very quiet on the subject. Production lines cannot remain open indefinitely while aircraft cannot go on being produced without prospective customers, and Concorde production facilities would have to make way for other work. By 1976 it was deemed that it was up to Air France and BA to make a success of the new supersonic era. However, without permission to fly to the USA, this appeared almost impossible.

On 4 February 1976, after protracted wrangling, the US Secretary of Transportation, William T Coleman, finally gave approval for BA and Air France to commence services for a 16-month trial period; one Concorde each per day to Washington and two each per day to New York. On 24 May 1976 Concorde services began to Dulles airport at Washington, owned by the Federal (US Government) authorities. In the words of Brian Calvert, Commander of that first flight: 'Planning started for what, it was decided, would be another spectacular – this time a joint arrival. We agreed that on this occasion the BA flight would land first – simultaneous landings were a little too much to expect.' The two Concordes performed perfectly, as they had done almost exactly four months earlier at the start of their commercial careers.

Opposite: A slender delta pivots around the slender CN tower. Concorde was a frequent visitor to the Toronto Airshow in Canada.

Overleaf: Meeting at Dulles tower. Scheduled Concorde services to Washington's Dulles airport began on 24 May 1976.

Landing Concorde at Dulles International Airport, Washington, was one thing, but permission to do so at Kennedy airport, New York, owned by the Port of New York Authority, was another. Coleman had implied that Federal pressure might be brought to bear on the Authority, but it became increasingly apparent that there would have to be a legal battle before Concorde could gain rights into New York.

Those rights were finally granted, and amidst threats of the greatest car-borne anti-SST demonstration, Concorde 201 (F-WTSB) arrived in New York on 19 October 1977. In command was Aerospatiale's Chief Test Pilot, Jean Franchi and on the flight deck with him were Captain Brian Walpole (BA Flight Technical Manager) and Captain Pierre Dudal of Air France. The following day with Captain Walpole in command, Concorde took off from runway 31L (the left hand of the two north-westerly facing runways at New York). For the first time Concorde made the famous left-hand climbing turn, started at 100ft above the runway. That turn, which was regarded as cheating the noise meters by some critics, had been the subject of much practice, both at other airfields and in the Concorde simulators. The turn, the maintenance of the correct speeds, the cutting back of the thrust, the reapplication, the cutting back and finally the re-application of thrust had been calculated precisely to correspond with Concorde's proximity to noise sensitive areas. The take-off was a success, the noise minimal. In one sense the protestors had won a great victory: Concorde had been made to be acceptably quiet and other airliners had better follow Concorde's example. On 22 November 1977 BA and Air France commenced supersonic services to New York.

Cartoons from London and New York newspapers comment on Concorde's struggle for New York landing rights.
Left: *The Times*, Friday 19 August 1977.
Below: The *New York Post*, Thursday 12 May 1977.

"Damn the court, DON'T LET IT LAND!!"

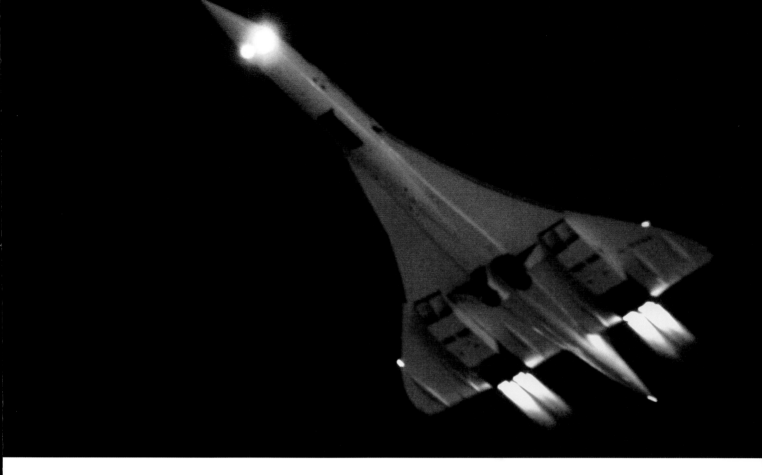

Best viewing of the reheats (or after burners) is achieved following a take-off in the dark. (Mark Wagner/aviation-images.com)

Had there been no delay in starting the New York service, there might have been further orders for Concorde. But however 'well dressed' Concorde had appeared, no purchasers were forthcoming while she had 'nowhere to go'.

In December 1977 BA, in conjunction with Singapore Airlines (SIA), extended the London to Bahrain Concorde service to Singapore. During performance and hot weather trials in September 1974 the first 'production' Concorde (G-BBDG), with Brian Trubshaw and Peter Baker (Assistant Chief test pilot BAC) at the controls, had flown supersonically over India. Supersonic overflying permission was now withdrawn so the Concorde route to Singapore had to go to the south of Sri Lanka adding some 200nm or ten minutes flying time. By December 1977 minor modifications and improved operating procedures

had increased Concorde's range, so, what would have been unattainable in early 1976, was by then quite possible. However, after three return flights the Malaysian government withdrew flying rights over the Straits of Malacca. After more negotiations with the Malaysians the route reopened in January 1979 and was operated until November 1980. Concorde was flown to Singapore by BA crews with the cabin alternatively being staffed by BA and SIA cabin crews.

World recession brought about a drop in Concorde's load figures between London and Singapore. This, coupled with the financial arrangements between BA and Singapore Airlines, put a greater share of the burden of loss onto BA, finally causing the route to be abandoned. A particular Concorde had been earmarked for use on this route (G-BOAD), being painted in Singapore livery on the left-hand side, and in BA colours on the right. This was the first and, by 1990, the only Concorde to have appeared in colours of an airline

G-BOAD showing Singapore Airlines colours on the port (left) side, taking off from Heathrow for the nine-hour journey to Singapore. (© Hulton-Deutsch Collection/Corbis)

other than those of Air France or BA. Sadly, as the route closed down, the BA flight deck crews returned home from their postings in Singapore. With an excess of crews, some would leave and others be redeployed in the airline. Morale which was usually very high, took a temporary dip. G-BOAD was restored to BA livery.

Earlier that same year the agreement between Air France, BA and Braniff which enabled Concorde to continue through to Dallas, Texas, from Washington (changing to an all American crew just for this sector), was wound up. The route lasted from January 1979 until June 1980. Braniff was, by 1996, the only American airline to have operated Concorde, albeit subsonically. They had hoped to gain experience with the aircraft for eventual supersonic services to South American destinations. In the meantime Concorde's high profile would have helped their marketing effort. But world recession and the deregulation of American

internal airline routes had stretched Braniff's resources, so their Concorde operation experienced unacceptably low loads. On the Washington to Dallas sector the Concordes bore their home airlines' livery, the only noticeable external change being the dropping of the 'G' in the registration. The British Concorde had been changed from the usual British registration of (G-BOAA) to a 'G' followed by an 'N', two single digit numbers and then the two final letters of the original registration letters (G-N94AA).

To give Concorde a boost during the time when services were declining, several Concorde crew members chartered the aircraft, giving people flights on Concorde at a fraction of the cost of a transatlantic fare. The first one to do so was a Concorde stewardess, Jeannette Hartley, who organised charters at no little financial risk to herself. In 1981 my wife and I chartered Concorde to celebrate Britain winning the Schneider Trophy contest. The last contest was won in 1931 by an RAF team captained by Squadron Leader AH Orlebar (a cousin of the author). In 1981 Concorde flew on two occasions, at

340mph, almost exactly 1,000mph less than her cruising speed, around the final course which was situated between the Isle of Wight and Portsmouth (England). The speed of 340mph was chosen since it had been the average speed of Flt Lt John Boothman's winning Supermarine S6B in 1931. Two Schneider trophy pilots flew on Concorde to celebrate the 50th anniversary of the final victory: Air Commodore D'Arcy Greig (1929 team) and Group Captain Leonard Snaith (1931 team). Other Concorde crew members followed suit.

On Concorde's financial side, a review in February 1979 of the ability of BA to make a profit with Concorde had concluded that it could not do so by ordinary commercial standards. Accordingly the Labour Government decided to write off the £160 million of Public Dividend Capital (PDC) associated with BA's acquisition of its five Concordes. In November 1979 an Industry and Trade Select Committee, chaired by Sir Donald Kaberry MP, was convened to investigate Concorde. Quoting from the report of their investigations (published in 1981): '… this [the writing off of £160 million PDC] meant that the whole fleet, including the initial inventory of spares, was entered in the BA balance sheet as a fully depreciated asset – that is to say a gift from the taxpayer. For their part BA had to pay to the Government 80% of future Concorde operating surpluses (the so called 80:20 agreement), though these were to be calculated after the offset of any operating deficits incurred after the review and also the amortised cost of any post-review expenditure.'

The BA partnership deal with SIA on the Singapore route negotiated by Gordon Davidson (Concorde Marketing Director, BA 1975–1979), appeared promising for BA, but it had dashed Government hopes of SIA purchasing at least one of the remaining unsold Concordes – two in Britain and three in France. Later the best return for the Government appeared to come from placing the two unsold British Concordes with British airlines. When Gordon Davidson moved to British Caledonian (BCal), rumours started of BCal operating a supersonic service to Lagos, but this came to nothing. For a period BA operated one of the two surplus Concordes when one of its Concordes was undergoing modifications. Ultimately BA acquired the two remaining Concordes, making seven in all.

According to the report, the UK had spent about £900 million on Concorde by the end of 1980. Furthermore the Government was still having to finance the support given by the manufacturers to the project, expected to total £123 million for the five years beginning 1980/81.

During the commercial life of an aircraft, its manufacturers not only undertake to supply spare parts, but just as importantly to supply a 'support' service in the form of carrying out development work and monitoring performance, suggesting or insisting, where necessary, on improvements or modifications. Overseeing this process are the aviation authorities in whose countries the aircraft are built and registered – the Civil Aviation Authority (CAA) in Britain. These aviation authorities have the full backing of their country's law behind them. Such support is usually financed out of the profits from the sales of spare parts and ancillaries to airlines operators.

Concorde was chartered on 12 and 13 September 1981, by the author to mark the 50th anniversary of the Schneider Trophy. Each of the passengers received a certificate. The Captain on those occasions was John Eames, the author was First Officer and Senior Engineer Officer David Macdonald was the Flight Engineer.

In the case of Concorde, monitoring of the aircraft's performance was, in the early days of operation, a very expensive business. Not least of those expenses were the two test specimens – full-sized Concorde fuselages dedicated to being tested on the ground. Quite early in the programme the stress test specimen in France had been purposely tested to destruction, but by 1980 the one at Farnborough was still costing several millions per year to run. This Concorde had been dedicated to being 'flight cycled', which included the heating and cooling process experienced on every supersonic sector. Profits through sales of spare parts would not raise anything like enough to pay for the Farnborough rig, nor enough to pay for the necessary, but diminishing, development work associated with Concorde. Accordingly, the four manufacturers – Aerospatiale, British Aerospace (successors to BAC), Rolls-Royce and SNECMA – were funded for their responsibilities to Concorde by their respective governments.

In October 1978 the jigs used to manufacture Concorde were removed from the Brabazon Hangars at Filton for storage at Wroughton, near Swindon in Wiltshire, against the possibility that there might be a demand for more Concordes. However, the French had, by December 1977, not only removed but disposed of their jigs; with their capability gone further production

Calshot Spit was the venue for the teams of the final two Schneider Trophy contests, those of 1929 and 1931. Here, in 1981, the Spit can be seen ahead and to the left of Concorde and was the venue for the 50th anniversary celebration of the final British win. A cousin of the author – Squadron Leader Harry Orlebar – was captain of the High Speed Flight of the RAF. RJ Mitchell who designed the winning S6b went on to design the Spitfire. Interestingly, a Spitfire was used to investigate the effects of diving at Mach 0.9+.

The Concorde Fatigue Specimen at RAE Farnborough. Although fully visible, here the airframe was festooned with ducts connected to a hot air supply in order to simulate the heating and cooling experienced by a Concorde on a supersonic flight. Each 'flight' was minutely monitored by computers. The experience gained by the time this specimen was dismantled in 1985 would have allowed Concorde to fly well into the 21st century.

of Concordes would have been an extra expense. On 31 December 1980 it was announced that the production phase of Concorde had ended, but it was not until October 1981 that disposal to scrap merchants of the stored British jigs began. About this time Federal Express investigated the use of Concorde as a supersonic parcel carrier, but this also came to nothing.

The Select Committee, charged to look into Concorde affairs, had to face some rather unpalatable truths. No more Concordes would be sold, the funding of the manufacturers would bring no return to the Governments, and the two airlines concerned did not seem able to operate Concorde profitably. It had been asked to make recommendations specifically on the question of costs due to be met out of the public purse. Was there, it tried to answer, a case for continued expenditure, which although unjustifiable in financial terms, might confer other benefits on the UK, such as prestige or the basis for starting a second generation SST? Then, more sinisterly, from Concorde's point of view, the Committee was to investigate how the cost of immediate cancellation would compare with the cost of continuing and how a proposal to cancel would affect relations between the UK and France.

The Committee heard evidence from representatives of BA, the British manufacturers of Concorde (British Aerospace and Rolls-Royce), the Minister of Trade and Industry (Norman Tebbit, an ex-BOAC pilot), and the

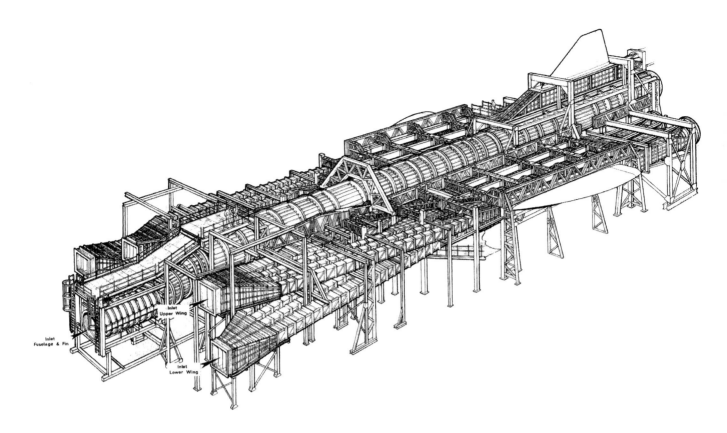

Above: The Concorde Fatigue Specimen engulfed by its thermal duct. The thinness of the wing tips and top of the fin meant that they would not suffer significant thermal fatigue, hence they were not under test. The fuel tanks were filled, the cabin pressurised and jacks were used to 'load' the airframe as on a real flight.

Below: The BA 195 being prepared for service at Heathrow on a winter's evening. Later, from the west, the sun will rise on this Concorde for the second time in the day. For a while during the early 1980s the livery excluded the word 'Airways'.

Deputy and Assistant Secretaries of the Department of Industry. Representing Concorde's paymasters, the Select Committee had every right to question the commercial decisions made by the companies responsible for servicing and operating Concorde.

It appeared that the writing was on the wall for Concorde, the more so since the return of a Conservative Government in May 1979. There is no doubt that they wished Concorde to continue if at all possible, but not at any price.

The change of government in Britain also saw a change of policy towards the nationalised industries. In Febuary 1981, Ross Stainton retired as Chairman of the BA Board, and Sir John King took over. He had been charged by the Government to prepare the company for privatisation. Having suffered disappointing financial results, attributable to the world recession, the moment had arrived to study every aspect of the airline with the view to cutting costs and increasing revenue. Nevertheless, the new leadership was more keen than the old on retaining Concorde.

Concorde was frequently chartered to destinations all over America. On this occasion, two weeks after the unveiling of the new livery, Concorde was used to publicise the opening of the Jumbo service between London and Pittsburgh via Washington.

All too frequently large organisations become conscious that their staff can lose a sense of identity and pride in their work. Following the period of recession, the restoration of morale in BA was regarded as a most important part in its march back to profitability. To this end, 'profit centres' were set up within the airline. One such centre was the Concorde Division. In May 1982, Captain Brian Walpole, who had been Flight Manager Technical was asked by Gerry Draper (BA Commercial Director) to become General Manager of this new division. As his assistant he brought with him another Concorde pilot, Senior First Officer WD (Jock) Lowe. Broadly they divided their responsibilities into two, Walpole looking into ways of increasing revenue and Lowe at ways of reducing costs. The appointment of practising pilots to these positions represented a welcome change from tradition. However, it must be stated that their predecessors in Commercial Division had worked very hard for Concorde's success in conditions which had not been easy. At that time Concorde passenger loads were falling from their peak in 1979 while costs consistently exceeded revenue.

Within months of their new appointment Concorde was faced by a new crisis. In August 1982 Ian Sproat MP wrote to Sir John King stating the Government's intention to cease funding the British manufacturers' support costs for Concorde and asking BA if it wished to take on this responsibility. If BA were to decline, in the words of Keith Wilkins (Head of Planning, BA): 'The supersonic project would terminate.' Termination in Britain at this stage would probably have meant termination in France as well. The date set for this was 31 March 1983.

President Mitterrand, unlike his predecessor, did countenance a review of Concorde's financial performance when he came to power in 1980 and Concorde was on the agenda of the Anglo-French summit of 1981. However, at no time in Concorde's history had Britain and France both shared the same opinion regarding cancellation, hence its survival. In Britain at this time, the Government perceived that by 1983 or 1984 the 80:20 agreement might even be producing a return from which the support costs could be financed, especially since the cancellation of the Singapore route. But Government involvement of this kind was not within the philosophy of the Conservative party in power. Probably for this reason, rather than from wishing to stop Concorde, did Ian Sproat write to Sir John King stating the Government's intention of ceasing to fund the British manufacturers.

The BA reply indicated that they were very willing to examine the possibility of taking over the support costs, but they would have to be given time to examine the implications. The Government agreed to another

Above: Washington Dulles, September 1985, passengers disembarking from G-BOAA via the 'Mobile Lounge'. Once loaded the lounge was lowered prior to being driven to the terminal building. After a 50-minute turnaround the aircraft will be en route to Miami.

Below: Concorde G-BOAA's inaugural service to Miami from London via Washington, 27 March 1984. The one hour 20 minute sector from Washington to Miami was flown at subsonic speeds over land, then at supersonic speeds over the sea south of Wilmington, North Carolina.

year being made available before the axe would finally fall on public money funding the British manufacturers. In the meantime a Department of Trade and Industry Review Group headed by Mr Bruce MacTavish of the Civil Service, would negotiate with BA for the handing over of the Government's responsibility to BA. The BA negotiating team were led by Mr Keith Wilkins (Head of Planning), with Captain Walpole (General Manager Concorde Division), Jock Lowe (Planning Manager Concorde), Sandy Sell (Engineering), and Peter Brass (Accounting).

The team's first job was to find out whether Concorde could make a sufficient operating surplus to fund the British manufacturers' support costs. Their second, and no less important, task was to analyse the manufacturers' activities with a view to reducing costs without impairing service, especially in areas of flight safety. This would rule out any development work not called for jointly by manufacturers and BA. The team were modestly optimistic that Concorde revenue which,

An Air France Concorde being prepared for service at Kennedy in October, 1985. By 1984 Air France Concorde operations became profitable on the Paris–New York route. Runway 22R was one of those preferred for a Concorde take-off on the basis of restricting noise (see page 82). There were no similar restrictions for Concordes landing at Kennedy.

had, in the period 1979–1980 to 1982–1983 been dropping, would improve as the recession passed. Over optimism was not only unwarranted, but might have upset the negotiations. Equally too much pessimism might have brought the negotiations to an untimely end. It took 18 months to find a satisfactory formula.

The main points at issue were the terms surrounding the acquisition by BA of Concorde spare parts owned by the Government (and useless to anyone other than a Concorde operator), the winding up of the 80:20 agreement soon expected to give government a small dividend, and the replacement of the Government by BA in the contracts with the British manufacturers (British Aerospace and Rolls-Royce). On this particular issue it was fundamental that BA could withdraw unilaterally from the Concorde project at its own discretion without having to finance the British

manufacturers if Air France continued Concorde operations. BA made it clear to the Government that they would not pay for the continuation of the operation of the test rig at Farnborough. This had been a major element in the support costs and by 1984 the full-scale Concorde structure had experienced sufficient 'flight cycles' for the Concordes to continue, at their present rate of use, well into the next century (see page 137). In the end BA took responsibility for the dismantling of the rig.

The funding of Concorde's French manufacturers would not necessarily be altered by the proposed changes in Britain – that was a French matter. Hitherto the French manufacturers had received payment from the French Government. In the early 1980s the French Government had promised Air France that it would bear a higher proportion of its Concorde operating loss. In return it reserved the right to dictate where Air France operated Concorde. The result was that the Rio de Janeiro, Caracas and Mexico through Washington routes were abandoned leaving Air France a single daily return Concorde service between Paris and New York. In 1984 Air France operations became profitable.

In April 1983, as discussions between BA and the Government continued, a 30-minute documentary programme about Concorde appeared in the QED series on BBC1. I was technical consultant to the producer, Brian Johnson, and was present as First Officer to Brian Walpole on the London to New York service which was filmed. (The Flight Engineer on that sector was Senior Engineer Officer Bill Johnstone.) Brian Walpole, during an interview on the programme, made it very clear that in future Concorde would have to stand on its own two feet: 'I believe it can, and, given a reasonable response from Government, Concorde will continue.' Although the tide had started to turn in favour of Concorde before the transmission of that programme, it was from that moment that Concorde ceased to be regarded as a loss maker. Gone were the tiresome yet familiar gibes about its poor prospects.

Slowly and inexorably the negotiations eroded the major outstanding differences between the Government and BA. In the end a sum had to be agreed which BA would pay to the Government for all the Concorde spares, Concorde G-BBDG (202) which was grounded at Filton minus engines and much equipment, the Farnborough Concorde test structure, as well as buying its way out of the 80:20 agreement. In March 1984, 18 months of detailed analysis were brought to a swift conclusion. In a meeting lasting not more than a quarter of an hour, reminiscent of bargaining in an eastern bazaar, Gordon Dunlop (Finance Director of BA) and the Government representatives agreed on a figure: £16.5 million. Concorde was saved.

Great credit is due to the people whose determination found a way of preserving this unique aircraft in service. In particular Bruce MacTavish as negotiator for the Government and Keith Wilkins, astute leader of the BA team, deserve great praise. Brian Walpole's and Jock Lowe's infectious enthusiasm and dedication to Concorde were great motivators throughout. During all the negotiations Concorde continued in service thanks to everyone connected with the operation, maintaining faith that somehow Concorde would have a successful future.

A Conservative Government spawned Concorde and ensured its right to life through an unbreakable treaty with France in 1962. Twenty-one years (and a few months) later, another Conservative Government severed almost all its financial connections with Concorde; the supersonic airliner had come of age. With expanding charter services and improving figures on the scheduled routes (London to New York and London to Washington and Miami), the commercial future of Concorde in BA looked bright.

THE FLIGHT – ACCELERATION

RECOLLECTIONS OF THE PREPARATIONS FOR SUPERSONIC FLIGHT:

'Ladies and gentlemen. This is the First Officer, Christopher Orlebar,' came my voice on the cabin address system. 'We are climbing through 20,000ft and are just accelerating through the cruising speed of a jumbo jet: Mach 0.85, 550 miles per hour. We are flying over the track of Brunel's Great Western Railway – another great engineering project – towards Bristol, and in particular Filton, from where this Concorde first flew in 1979.

'The time in New York, if you would like to reset your watches is five minutes to six. We expect to arrive in New York at a quarter past nine. Having left London at ten thirty, you do not need to be an Einsteinean physicist to work out that that will make us one and a quarter hours younger by the time we arrive at Kennedy airport. Although we travel backwards in time it is not sufficiently far to call our destination by any of its original names – Idlewild, or New Amsterdam.

'In six minutes from now we shall be switching on the after-burners. You will feel two small nudges as they come on in pairs. They give extra thrust to the engines to overcome the increased air resistance found during supersonic flight. We accelerate and climb, and by 43,500ft we shall have achieved Mach 1.7, at which

point the after-burners will be switched off. In case there are no thermodynamicists among you I will explain why. By Mach 1.7 the engines will have become very much more efficient due to the increase in airflow through the intakes, which precompresses and slows down the air before it enters the engines. There is then sufficient thrust to overcome the increased air resistance caused by those shock waves which appeared just below Mach 1. We continue to climb and accelerate, reaching Mach 2 just above 50,000ft. Thereafter we climb gently as we use the fuel and so become lighter, maintaining Mach 2, or thereabouts, until we reach today's ceiling of about 58,000ft – in any case not above 60,000ft which is as high as we are allowed to fly.'

Another voice now, this time in the headsets of the crew, from the air traffic controller at West Drayton: 'Speedbird Concorde one nine three, London. You are cleared to climb at the acceleration point. Cross eight degrees west at or above flight level four three zero (43,000ft).'

'Roger. Cleared to climb at the acceleration point. Cross eight degrees west flight level four three zero or above. Speedbird Concorde one nine three,' comes the reply to London Air Traffic Control. Soon Concorde will be travelling at almost two and a half times the speed of a jumbo jet.

Two hours previously the three flight deck crew of BA 193[*] assembled in the Queen's building at London's Heathrow Airport.

Senior Flight Engineer Tony Brown (left), Captain John Massie (centre), and the author at the flight briefing in the Queen's Building at Heathrow. In 1994 British Airways moved its Flight Crew Briefing Centre from the Queen's Building in the central area to the Compass Centre on the north side of the airport. Concorde never operated from Terminal 5.

'One hundred [passengers] booked. You have Alpha Foxtrot on stand Juliet 2.' Armed with this information and a list of the six cabin crew members, the three crew, consisting of Captain, First Officer and Flight Engineer, go to be briefed. They study the weather, the fuel flight plan and other relevant information concerning airfields, navigation aids and route information.

The weather forecast at the destination and alternate airfields is of fundamental significance. Is fog likely? Concorde is fitted with a Category Three automatic landing system, which allows landings in visibilities down to 656ft (200m) with a 'decision height' (whether to land or to go up again) of 15ft (4.5m). Whatever the weather at the destination, the aircraft must be able to divert to an alternate airfield whose weather would allow an ordinary manual landing. The weather at the en route alternate airfields is also checked. These might be needed in the event of engine failure when Concorde would be forced to decelerate to subsonic speeds where the range is not so great.

The forecast winds between 50,000 and 60,000ft are studied. They are usually less strong than the winds

found up to 40,000ft. Very rarely do they exceed 100 knots. On average they blow from the west at 30 knots. Cruising at 1,150 knots true airspeed, a 100 knot headwind has less effect on Concorde than on a jumbo which cruises at 480 knots. Over a 2,000nm distance a 100 knot headwind gives Concorde a ground speed of 1,050 knots increasing the flight time by ten minutes, but a jumbo with a ground speed of 380 knots would take 65 minutes longer over the same distance. The tracks followed by subsonic aircraft over the Atlantic are varied from day to day to take account of the winds. The supersonic tracks, however, are fixed by the minimum distance consistent with over-water flight where the sonic boom is acceptable.

Passengers at the check-in desks for BA001 to New York. From here passengers would be fast-tracked through to the Concorde lounge. (Mark Wagner/aviation-images.com)

* In 1987 the flight number of BA 193 was changed to BA 001.

Concorde G-BOAA during refuelling on stand 'Juliet 2' at London Heathrow, Terminal 3. A maximum of about 95 tonnes, depending on the density of the fuel, is distributed around the 13 fuel tanks. The bowser simply pumps the fuel from a distribution point beneath the parking stand to the aircraft, measuring the quantity as it is loaded. The metallic plates beneath the front left-hand door on the aircraft house the 'static pressure ports'. The atmospheric pressure sensed here is fed to an Air Data Computer which amongst other things sends altitude information to the altimeters. The device around the nose wheel tyres is there to deflect water and slush from the runway away from the engine intakes; the main gear has a similar device. Concorde transatlantic operations from Heathrow were moved to Terminal 4 in 1986.

Concorde G-BOAF at Terminal 4, Heathrow. The 'elevons' are drooped since, without the engines running, there is no hydraulic pressure. During the prestart checks the electric power must be supplied by an external source; the weight penalty of an Auxiliary Power Unit would have been excessive. Ground air-conditioning and air pressure for engine start had also to be from external sources. Note the tail wheel, fitted to avoid the exhaust nozzles of the engines contacting the ground on take-off and landing following an 'over-rotation' and an excessive angle of bank. The prototypes were fitted with a tail bumper.

The temperature to be found at altitude is also of importance. The average over the Atlantic constantly varies, but is of the order of –55°C (–67°F) at Concorde's cruising level. A few degrees warmer and the fuel requirement is greater and vice versa. For efficiency the engines prefer cold air. There is a greater mass of oxygen in a given volume of air at a given pressure, in cold air than in hot air.

The fuel required to carry 10 tonnes of payload (100 passengers and their luggage) and 15 tonnes of spare fuel over the 3,150nm between London and New York is typically 77 tonnes (depending on winds and temperatures). Nevertheless the fuel flight plan is meticulously checked. For taxiing at London 1.4 tonnes are added bringing the total to 93.4 tonnes – some 2 tonnes short of full tanks. The expected take-off weight with these figures is 180 tonnes. The expected flight time on this journey is three hours and 23 minutes.

Equipped with the paperwork the crew are driven to the aircraft on stand Juliet 2. At this stage the gleaming paintwork showing off the new BA Concorde livery (unveiled 25 April 1985) is somewhat hidden by hordes of service vehicles, fuel bowsers, luggage and catering trucks and transport vans. The passengers congregating in the special Concorde lounge glimpse the unique supersonic nose serene above the white-overalled activity beneath.

It is almost impossible for the passenger, viewing all this activity, to imagine just how many lifetimes of thought and effort preceded this moment of sublime anticipation: flying the world's only successful supersonic airliner. Yet as these words were written, powered flight by men in heavier than air machines has happened within the life of many living people and commercial supersonic flight was already ten years old.[*]

Overleaf: G-BOAC's cockpit. Unpowered, the instruments display red failure flags. (Mark Wagner/aviation-images.com)

Captain Tony Meadows, one of the original BA Concorde pilots, carrying out the pre-flight 'scan' checks. Each item is checked one after another following a strict pattern. Designed in the early 1970s, the instruments on Concorde are all electro-mechanical in contrast to the screens showing computer-generated displays which began to feature in the 1980s.

The 'safety' checks having been completed in the cockpit, the pilots remain on board while the Flight Engineer checks the exterior of the aircraft. He will inspect, amongst other things, landing gear, tyres, engine intakes and the elevons to the rear of the wing, which look, without hydraulic pressure, rather like flaps in the 'down' position. As he does so he steps over power cables, refuelling hoses and the high-pressure air hoses for use during engine starting. Finally he rejoins the pilots going through their 'scan' checks. Lights, instruments and audio warnings are all scrutinised. 'Pull up, pull up!' says an insistent microchip from the loudspeakers. It is the terrain avoidance system under test. Another voice, taped and transmitted from the tower, announces: 'This is Heathrow information Romeo, zero eight one five weather, wind two nine zero degrees, one five knots...' The data is copied down, relevant for calculating whether conditions are suitable to allow 180 tonnes of Concorde to take off from runway '28 Left' (the southern of the two west facing runways at Heathrow).*

Although more crowded with instruments than other airliners, the cockpit in Concorde is equipped with the familiar ones: altimeters, airspeed indicators, artificial horizons and engine temperature and pressure instruments. However, the Mach-meters are calibrated up to Mach 2.4, and on closer inspection switches and gauges of systems unique to Concorde become apparent – engine air intake controls, primary and secondary engine nozzle indicators and a centre of gravity meter – but despite diligent searching, there are no flap and slat levers. To the untutored eye it is as foreign and daunting as a cathedral organ is to the tone deaf. But to the crew it is like home; a place that both gives and demands care and nurture.

'Good morning gentlemen.' It is the Cabin Services director introducing himself to the flight deck crew. The cabin, with the new 'space-age grey' upholstery, has been meticulously checked by the six cabin crew, as has the food and drink to be consumed during the flight. However well the aircraft is operated technically, it is the cabin crew who must supply the ambiance and good feeling that will make the passengers say: 'We had such a good flight...' Dedicated to Concorde they meet the challenge. They are lucky since they can see the fruits of their dedication – satisfied passengers. But members of the cabin crew are trained in other equally vital tasks affecting the well-being of the passengers; they must be experts with all the emergency and survival equipment that is carried on board.

No less dedicated, but behind the scenes, were the thousands of individuals who made the flight possible. The shifts of engineers, the refuellers, the catering staff, the Chairman, the ramp controller, the Chief Executive, the tractor driver, the administrators... the list goes on. Their roles were vital, their aim identical – the successful flight of BA aircraft. That Concorde performed so well

* Rereading these paragraphs in August 2010 it seems remarkable that it is now seven years since the end of Concorde's flying days. Progress is usually associated with faster communication or transport and yet we are now no further forward than when Concorde first took to the air.

* Since July 1987 this runway has become '27 Left', because the magnetic variation has reduced.

A watercolour by Roy Huxley of Concorde on a taxiway. Based on a picture taken by the author with a wide-angle lens, the wings appear more swept back than is the case. The nose and visor are at 5 degrees ready for take-off.

is proof that great pride in the job existed with this section of BA. Success breeds success; the 'Halo' effect of Concorde benefitted all sections of the airline.

Familiar calls can now be heard.

'OK for boarding?'

'Fuel book,'

'Number four uplifted four quarts,'

'No performance A.D.D.'s, but there is a history on number two H.F. radio, it is very slow to tune.'

'Speedbird Concorde one nine three, cleared to New York Brecon one foxtrot, squawk two one zero three, track Sierra Mike.'*

'Cleared to start, one two one nine for push'.

The clearance is read back by the crew to the tower.

'Fifty three and half for take-off with a burn-off of 900 kilograms,' comes another statement. It is akin to the tuning up of an orchestra.

The Captain or First Officer may fly the aircraft. Today it is the First Officer's turn, so the Captain acts as co-pilot to the First Officer although he retains overall command. Thus the Captain reads the 'before start check list'. Henceforward all actions are coordinated between the three members crew as a team, and each one, where possible, monitors the others' actions.

'Altimeters,' says the Captain.

'QNH one zero one five set, eighty, six eighty, normal and 20ft on the radio altimeter,' comes the reply. Later: 'INS one, two and three'. Concorde is navigated by three inertial navigation systems (INS). Once programmed with the latitude and longitude of the starting point of the aircraft, the INS by sensing acceleration and referencing it to the three dimensions (with a correction applied to take account of the earth's spin), will display the aircraft's position. Programmed with a route (a series of turning or 'way-points') the autopilot will steer the aircraft along that route. All three computers are checked.

'ASI bugs and pitch indices,' says the Captain. 'Vee one, one six zero; rotate, one nine five…' These are the relevant take-off speeds. The first one, 160 knots, is the decision speed. Up to this speed the aircraft may safely stop within the runway distance; beyond, the take-off may be continued safely even with one of the four engines out of action. On this journey the aircraft will be 'rotated' (that is pitched up) to 13 degrees at 195 knots (225mph) to become airborne at 217 knots (250mph).

The loadsheet is checked. 90 of the 100 passengers have turned up and are on board; there are ten 'no-shows'. The zero fuel weight and the zero fuel centre of gravity have been programmed into the relevant

* Track 'SM' is the westbound Atlantic supersonic track (see endpapers).

computer. They are finally checked. Fuel is shifted or 'burnt off' prior to take-off to position the centre of gravity to one specific point. With no tailplane, the precise position of the centre of gravity on Concorde is more important than for an aircraft with a tailplane.

Finally: 'Start engines'. The inboards are started first. All four on, even at idle power, might damage the towbar on the tractor during the push-back. Hydraulic pressure from pumps driven by the engines at 4,000psi is fed to the flying controls. They spring into life, ready for the comprehensive 'flying control check' that follows. The hydraulic jacks are signalled electrically, but have a mechanical back-up channel. The electrical signals are modified by an auto-stabilisation system which reacts similarly to the reflex actions in a human. The outboard engines are started, the tractor and towbar unhitched, communication with the ground engineer is cast off and Concorde taxis towards runway 28 Left.

The view from the left-hand seat. It has been said that glass is the one man made commodity that has transformed civilisation more than anything else. Storing fluids, allowing light in and keeping rain out,, making lenses, spectacles, telescopes and microscopes; then in the modern world electric light, thermionic valves (that preceded transistors), televisions, computers and here on Concorde the visor acts as a streamlined heat shield with the advantage of being transparent.

'Good morning, ladies and gentlemen. This is the First Officer adding my words of welcome to those of the Captain. I would like to describe the take-off, with an apology to those of you who already know our procedures so well. There is a greater thrust to weight ratio on Concorde than there is on subsonic aircraft so the take-off is a bit more sporty, if I can put it that way. One minute and 20 seconds after the start of the take-off run we shall be reducing the thrust and switching off the after-burners, which have been adding to that thrust. This reduces the noise near the airport. Inside the cabin you will notice both a reduction in noise and in the angle of climb. By about 12 minutes after take-off we shall be flying at 95% of the speed of sound – Mach 0.95. Once we clear the coast of South Wales we shall accelerate to Mach 2. The weather in New York is perfectly satisfactory for aviators but not good for sunbathers – it is raining, but the forecast is for it to clear by our time of arrival.'

'Speedbird Concorde one nine three cleared for take-off runway two eight left,' transmits the controller from the tower at Heathrow. 'Roger cleared for take-off, Speedbird Concorde one nine three,' replies the Captain.

All the checks are complete.

'Three, two, one, now.'

On the 'now' the throttles are opened and the stopwatches started.

'Speed building,' calls the Captain, then: 'One hundred knots.'

'Power checks,' responds the engineer. Each engine is using fuel at over 20 tonnes per hour and giving 38,000lbs of thrust with afterburner.

'Vee one'. The First Officer moves his hand from the throttles, to the control column. There is no stopping now.

'Rotate,' the control column comes back and the whole aircraft rotates to an angle of 13 degrees above the horizontal. Concorde becomes airborne at 217 knots.

'Vee two' – a safe climbing speed in event of engine failure.

G-BOAF above the clouds, photographed in what was to be the BA Concordes' final livery.

'Positive rate of climb.' Now there is at least 20ft between the wheels and the ground.

'Gear up,' commands the First Officer. The Captain selects it up.

'Two forty knots', the pitch attitude is raised from 13 degrees to nearly 20 degrees to maintain 250 knots. Then looking to the stopwatch the Captain calls 'Three, two, one, noise.'

On the word 'noise' the engineer switches the afterburners off and adjusts the throttles to a preset position on the throttle quadrant. To maintain 250 knots with less thrust the angle of climb must be reduced. The First Officer pushes the control column gently forward to maintain a new attitude of 12 degrees to the horizontal. The rate of climb reduces from 4,000 to 1,200ft per minute.

At 7nm from London more power is applied and the speed is allowed to rise consistent with crossing a radio beacon called Woodley (close to Reading) at 4,000ft or higher. Accelerating out of 250 knots the nose and visor are raised and as the view of the outside world diminishes, a glorious calm settles on the flight deck. By 300 knots the flight becomes smoother as the buffet from the 'vortex' lift disappears. There are no flaps

and slats on Concorde, the same wing-shape serves throughout Concorde's 1,160 knot speed range.

'Speedbird Concorde one nine three, zero, climb and maintain flight level two eight zero,' says the London Air Traffic Controller.

The clearance to climb is acknowledged by the Captain; 28,000ft is programmed into the autopilot. The indicated airspeed during the climb is 400 knots. It is also as fast as the aircraft is allowed to fly between 6,000 and 32,000ft, the limit being shown by a black and orange chequered pointer on the airspeed indicator. Climbing at a constant indicated airspeed into the thinner air means that the Mach number will rise. At 25,000ft it will have risen to Mach 0.93 – the limit for a subsonic climb. At that point there will still be about 70nm to go before the coastline – the acceleration point. Once clear of the coastline, Concorde can accelerate to a Mach number which would cause the sonic boom to be heard. Usually the boom does not reach the surface until Mach 1.15 is achieved, a figure dependent on temperatures and wind.

'Centre of gravity steady at fifty-five per cent,' the Engineer announces. As the Mach number builds the centre of lift starts to move to the rear. This has to be compensated for by the rearward movement of the centre of gravity by about 2ft, achieved by shifting the fuel. By the time Mach 2 is reached the centre of gravity will have been moved aft by a further 4ft.

The autopilot 'acquires' 28,000ft. As it does so the auto-throttles switch in and take responsibility for maintaining the speed – Mach 0.95 (385 knots indicated airspeed at this altitude), 100mph faster than most subsonic aircraft.

'Speedbird Concorde one nine three cleared cruise climb, flight level four nine zero to six zero zero.' The message is acknowledged and 60,000ft is programmed into the autopilot.

'Checks complete down to the after-burners; fuel is going aft; one mile to go,' says the Engineer. The First Officer pushes the throttle fully forward. A signal is sent to the engines, via their controlling computers, to give maximum climb power. The distance 'to go' shows zero.

'Inboard reheats…' a nudge as they light up increasing the thrust by about 20%.

'Outboard reheats…' another nudge. Each engine is now burning fuel at over 11 tonnes per hour. The Mach number climbs and hovers on Mach 1. The shock wave passes the static pressure-ports on the side of the fuselage causing a fluctuation on the pressure-driven instruments, notably on the vertical speed indicator. Mach 1.01 is indicated and Concorde is supersonic.

At Mach 1.3 the variable ramps inside the engine air intakes begin to operate. They arrange the shock waves formed in the intake mouth in the most efficient pattern possible to compress and slow the airflow down prior to its entering the engine face.

The acceleration becomes more rapid as the Mach number builds. The passengers watch the progress on the 'Marilake' indicators at the front of each cabin. At Mach 1.7 a barely perceptible lurch indicates that the afterburners have been switched off. Reference to the cabin indicators shows the aircraft to be climbing through 43,500ft, the outside air temperature to be –52°C (–62°F) and the ground speed 1,120mph. There is a 20mph headwind. Acceleration is now less rapid. 40 minutes from take-off Mach 2 is attained at an altitude of 50,200ft and lunch is served.

'Ladies and gentlemen, at the risk of interrupting the Marriage of Figaro on the in-flight entertainment – or worse still, your conversation,' says the First Officer, 'We are cruising at Mach 2 (a mile every 2¾ seconds) and climbing gently towards twice the height of Mount Everest – 58,000ft. Here at the threshold of space, the sky above is far darker, almost black and the view of the Earth's horizon just betrays the Earth's curvature. Today it is very clear but big volcanic eruptions in any part of the world, throwing up tons of minute fragments of debris into the upper atmosphere, can reduce the clarity.

'The sun is now climbing from the west. In winter it is possible to leave London after sunset, on the evening Concorde for New York, and watch the sun rise out of the west. Flying at Mach 2 in an easterly direction at these latitudes will cause the sun to set in the west at three times its normal rate, casting, as it does so, a vast curved shadow of the Earth, up and ahead of the aircraft.'[*]

Two hours out of London, Newfoundland is visible on the right-hand side. The passengers visit the flight deck. Foreign secretaries, famous people, chief executives, popstars, owners of publishing empires, financiers and Concorde admirers, who have just come to experience 20th century air travel at its most supreme, are among the passengers. All are treated as VIPs.

Some are stunned into silence. Then, 'Do you know what each of these switches and dials do?'

'No,' replies the First Officer, 'We only have them to preserve the mystique.' Laughter, then:

'Does Concorde really grow eight inches during the cruise?'

The engineer explains that, due to the compression and friction of the air, the temperature of the outer surface of the fuselage rises to about 100°C (212°F), hence the expansion and increase in length. He puts

---

[*] Mercifully the passengers did not choose Concorde solely to observe astronomical phenomena whilst eating haute cuisine served by the dedicated cabin crew. They flew on Concorde because it saved them days, not hours. Westbound, the critical working hours of the day are preserved; eastbound, the purgatory of the overnight sector is avoided. Concorde only flew between London and the United States during the waking hours of the Atlantic seaboard dwellers.

The initial cruising altitude out of London is 28,000ft. Note Mach number: Mach 0.95. The compass is referenced to TRUE north for use during the transoceanic cruise, as opposed to MAG (magnetic) north, which is used on departing and arriving at an aerodrome. Note the drift about 7 degrees right with a strong southerly wind.

his hand into a gap between his panel and bulkhead; it fits. 'Once the fuselage is cool there is no space to do that. If you want two flights on Concorde leave your hand there during the deceleration, it will become trapped. The only way then of removing it would be to wait for the next supersonic flight. It would be painful, but it might be worth it!' he adds with a grin.

Should the temperature on the nose, the hottest point, be about to exceed 127°C (260°F), then the Mach number has to be reduced. This occurs when the outside air temperature becomes warmer than -50°C (-58°F). The speed of sound is greater in warmer air so the reduced Mach number has little effect on the flight time.

'Boston this is Speedbird Concorde one nine three heavy flight level five six eight (56,800ft).' A small

cross with BA 193, FL 568 and 999 appears on the controller's radar screen. The 999 refers to Concorde's ground speed, the radar is calibrated no higher, so cannot show the 1,120 knots registered in the cockpit.

'Speedbird Concorde one nine three heavy. Roger. I have you radar identified. Omit position reports.' (The suffix 'heavy' serves to differentiate groups of aircraft, on the basis of maximum allowable take-off weight.)

Concorde is passing the south-western end of Nova Scotia now. The track is being precisely steered by the autopilot following instructions from the inertial navigation system (INS). A small yellow light captioned 'R Nav' illuminates in the top right-hand corner of the Captain's instrument panel. Nantucket DME (distance measuring equipment – a radio pulse beacon), over 250nm away, has just taken over the refinement of the almost impeccable accuracy of the INS. Travelling at 1,900ft per second the position is known to within about 2,000ft.

The end of the supersonic cruise is near.

THE FLIGHT – DECELERATION

SCENES FROM A SUPERSONIC JOURNEY

'Speedbird Concorde one nine three heavy, Boston, cross three nine four three north seven one zero seven west at flight level five two zero or above, cleared to cross Linnd flight level three nine zero or above. Descend and maintain flight level three nine zero.' The descent clearance is read back. Concorde must cross the boundary of a 'warning' area, dedicated to military flying, above 52,000ft, before entering the regular airways system at a point over the ocean called Linnd at 39,000ft or above, but no lower – yet.

The deceleration and descent distance is calculated; typically 120nm are covered from Mach 2 to Mach 1. Decelerating over the ocean en route to New York, the point at which the deceleration is started is dictated by the distance required for nearly the whole descent. On most other routes it is fixed by the need to be subsonic 35nm before any coastline sensitive to sonic booms; on those occasions Concorde decelerates through Mach 1 level at 41,000ft; into New York Concorde becomes subsonic during the descent through 35,000ft.

There is a sharp noise: it is an electronic rendering of a 'wake-up!' bugle call. The crew is carrying out the 'Deceleration and Descent Checklist'. The first item is to test the warnings associated with the autopilot disconnecting itself – hence the alerting audio warning. Next to be selected are the radio aids required for the landing and the procedure for climbing out again, if the landing is aborted for any reason. The altitude to which it is safe to descend and the whereabouts of the high ground or buildings are scrutinised. Finally the three crew adjust their safety harnesses.

'Ladies and gentlemen, in 200 miles – 10 minutes in Concorde, but half an hour for the Boeing 747 in the 100 knot headwind which it is experiencing beneath us – we shall be commencing our deceleration. The time in New York is twenty past eight in the morning. We shall be landing at five past nine, on runway four right – the right hand of the two north-east facing runways. The rain has cleared and the temperature is 70°F.'

Thrust is reduced to an intermediate setting. Too low a value might cause the engines to surge (backfire) – alarming, but not dangerous; also the racks of computers would overheat if the air conditioning supplied by the engines failed to supply a sufficient cooling draught.

There is a whiff of ozone. Between 40,000 and 60,000ft Concorde has been flying in an ozone-rich atmosphere. At 'cruise power' the compression of the air in the engines heats the air to over 400°C (750°F). This converts all the ozone present in the compressor into oxygen. The air bled from the engines prior to its being cooled in the air conditioning system is thus free of ozone. At reduced power, the compression and heating are less – hence the trace of ozone. In spite of

The view through the visor at speeds greater than 250kts. Once the angle of attack is less than 9°, vision ahead is no longer obscured by the nose and visor in the fully up position, as shown here. The view is excellent in comparison to the view through the periscopic device fitted to the prototypes.

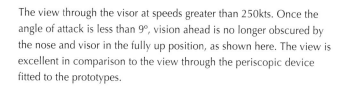

The author on the flight deck of Concorde 002, now at Yeovilton, showing the original visor/heat shield in the 'up' position. Even with a periscopic device to improve forward vision, the US Federal Aviation Administration would not have certificated this arrangement. Concorde subsequently benefited from improved glazing techniques.

the huge temperature drop in the supply of air to the air conditioning system, there is no temperature variation in the cabin.

At Mach 1.6 there is a further thrust reduction. Once the indicated airspeed has dropped to 350 knots the descent commences: 350 knots at 58,000ft corresponds to a Mach number of 1.55 (890 knots true airspeed); by 35,000ft it will correspond to Mach 1.[*]

The engineer moves the centre of gravity forward to keep the aircraft in trim. He pumps fuel from a tank in the rear of the aircraft to tanks in the fuselage and wings. Eight or nine tonnes of fuel are moved forward at this stage. The gap between his panel and the bulkhead closes and the warmth radiating from the windows reduces as the fuselage cools.

Further descent clearance is acknowledged: 'Roger New York, cross three five miles south-east of Sates at one two thousand feet, altimeter two nine eight four.' The altimeters are adjusted to show altitude referenced to the pressure at sea level, 29.84 inches of mercury. Hitherto, they have shown altitude referenced to a standard pressure of 29.92 inches of

mercury (1,013.25 millibars) – hence the instruction to maintain 'flight levels'. A flight level of 350 corresponds to 35,000ft with the altimeter referenced to the standard pressure setting of 1,013.25 millibars.

'Concorde one nine three, traffic 11 o'clock five miles south-west bound – one one thousand feet.' The crew peer slightly left and below. 'Contact, Speedbird Concorde one nine three.' 'Contact' means that the other aircraft has been seen – nothing worse. Although air traffic controllers are primarily responsible for the separation of aircraft under their charge, the crew includes a careful lookout as part of their cockpit routine. The view ahead, although the visor is raised, is surprisingly good.

As the speed falls the angle of attack must be increased to maintain the lift. 'Angle of attack' is not some kind of refined military manoeuvre; it is the angle between the aircraft and the oncoming air (see page 12). As it is increased so the view of the airspace directly ahead of the pilots becomes progressively obscured by the nose. By 250 knots the visor and nose must be lowered for adequate vision ahead to be maintained. Initially the nose is lowered to 5 degrees. On final approach, where the angle of attack is

14 degrees, the nose is lowered to its fully down position (12.5 degrees).

'All secure aft,' the chief steward reports. This means that he has checked that all the passengers are strapped in and the cabin equipment is stowed ready for landing.

'Ladies and gentlemen, this is the First Officer. We are under the control of the New York radar controller and will be landing in eight minutes from now.'

'Speedbird Concorde one nine three, fly heading three five zero degrees to establish on the localiser runway four right, maintain 2,000ft until established, reduce to two one zero knots.' The instruction is acknowledged, and the aircraft turns while 210 knots is dialled into the 'auto-throttles' and the 'IAS acq' button is pressed. As the speed falls below 220 knots the drag increases, calling for a thrust increase. Concorde is unique among airliners in flying at speeds below its minimum drag speed.

A short 'bugle call' is heard. The First Officer has disconnected the autopilot. Today he will carry out a manual landing. 'Localiser active,' calls the Captain. The radio beam defining the path to follow to the runway has been approached. The aircraft must be turned to the right from its heading of 350°M (referenced to Magnetic north) on to one that will let

Top: The three crew members at work. In the foreground is Senior Flight Engineer George Floyd.

1 Visor and nose in up position.

2 Visor down with nose at 5 degrees. The 'pitot' tube on the tip of the nose senses 'dynamic' pressure to measure airspeed. To the rear of a second pitot tube, halfway along the nose, is an incidence vane to measure, after compensation for the droop angle of the nose, the 'angle of attack'.

3 Nose at 12½ degrees, this position is used for landing. On a 3 degree glideslope with an angle of attack of 14 degrees the pitch attitude is 11 degrees, so the nose at 12½ degrees does not obscure the runway. If the nose failed in the 'up' position one option was to carry out an autoland. Should that not be possible a curved approach could be flown by the pilot on the downwind side of the aircraft. The 'direct vision' windows were not supposed to be open in flight.

Service at Mach 2. Concorde has a cabin crew of six, here shown in the uniforms of Concorde's tenth anniversary (**top**) and the Paul Costelloe-designed summer uniform (**above**) introduced in 1994.

Below: Haute cuisine at grande vitesse.

it track precisely along the radio beam transmitted from the ground. There is a crosswind from the right so the aircraft heads 48°M (3 degrees east of north-east) to track 43°M along the localiser beam of runway 4 R.

By 12 miles to touchdown the speed has been stabilised at 190 knots. 'Glideslope active,' calls the Captain. There are 9 miles to touch down. 'Gear down and landing check-list please,' calls the First Officer. The engineer reads the checklist. There is a double thump as the main undercarriage legs lock down, almost simultaneously. Four green lights appear confirming that all four have been extended – two main, one nose and a tail gear.

'Nose,' says the engineer.

'Down and green,' replies the Captain, referring, not to the colour of the nose, but to a green light which confirms that the nose is locked down. The view of the runway is now quite unobscured.

'Glideslope engaged,' calls the Captain. Concorde will now follow a second radio beam, this one slopes towards the landing point making an angle of 3 degrees with the horizontal. Together the two radio beams, glideslope and localiser, are called the Instrument Landing System (ILS). The word 'glide' here is a misnomer. No jet aircraft could glide at such a shallow angle to the horizontal with all the drag from its flaps, least of all Concorde, with the high 'induced' drag found when the slender delta wings are asked to give lift at slow speed. It is a powered approach. A speed of 190 knots is maintained through use of the auto-throttles.

The First Officer follows two yellow bars which form a cross over his artificial horizon. The horizontal one commands him to pitch the aircraft either up or down to maintain the 'glideslope'; the vertical one, to turn

Overleaf: G-BOAC. The 'fineness ratio' (length to diameter ratio) must be high to reduce 'wave drag' (caused by the formation of shock waves). Concorde's elegance was due to aerodynamic necessity. Since ceasing services Concorde has been missed, one comment being that 'now we have to set our watches by the BBC'. (John Dibbs/aviation-images.com)

The Flight Engineer's panel showing the fuel management system. In the centre there is the centre of gravity meter showing 58.8% (a measure of its position with respect to the Aerodynamic Root Reference Chord). Halfway along the chord would be 50%; 1% corresponds to about 1ft. The white pointer must lie within the two orange limit pointers (marked FWD and AFT) in order to maintain aerodynamic balance. The tank at the rear has 9,810kg of fuel in it. This will be moved forward to central tanks during the deceleration. The total fuel on board is showing 34,000kg. In this picture each engine has used about 15,000kg.

the height of the aircraft above the ground that the aircraft performs incredibly smooth automatic landings, somewhat to the chagrin of the pilots.

'Beep, Beep, BEEP, BEEP, Beep, Beep' in the headsets.

'Marker, height checks,' says the Captain. This is in response to this audible radio beacon being overflown at the correct height – 920ft. The veracity of the glideslope has been checked – a little academic with the runway so clearly in view, but vital in conditions of low visibility.

'Eight hundred feet radio,' calls the Flight Engineer. Now the speed is reduced from 190 knots to the speed required just over the threshold of the runway, 163 knots at a landing weight of 103 tonnes. Concorde decelerates during the descent from 800ft (2.34nm before touchdown) to 500ft (1.34nm before touchdown). To maintain 163 knots, more power is required than was needed at 190 knots. The procedure is called the 'reduced noise approach', because less thrust is needed throughout the approach down to 500ft. First because the drag is less at 190 knots than at 163 knots and secondly because less thrust is required during the period of deceleration. The noise generated by the engines is less with less thrust, and the fuel consumption is lower. Had an automatic landing been carried out, the final approach speed would have been achieved by 1,200ft above the surface, so stabilising the descent rate for a longer period to allow the landing computer to make a smooth touchdown.

During final approach Concorde consumed fuel at ten times the rate per mile than was the case towards the latter end of the supersonic cruise. At the intermediate speeds the fuel consumption was also much higher; hence the requirement to arrive with 15 tonnes of spare fuel, enough for 50 minutes in the stacking pattern prior to a landing.

'Speedbird Concorde one nine three heavy, cleared to land four right wind zero seven zero at one five knots,' says the controller in the Kennedy tower. Rarely is Concorde held up by congestion at this hour in the morning.

either left or right to follow the 'localiser'. The bars are signalled by the automatic flight control system. Had the autopilot been engaged it would have pitched and turned the aircraft to follow the two radio beams.

'One thousand foot radio,' calls the Flight Engineer. This is the height of the main wheels above the surface, determined by bouncing a radio wave off the surface. It is a far more precise measurement of height than that available through the pressure altimeter; however, it only works below 2,500ft. So well does it measure

'Cleared to land Speedbird Concorde one nine three heavy,' acknowledges the Captain.

'Five hundred feet,' calls the Flight Engineer. It is somewhat reminiscent of a 19th century sailor calling out the depths he has plumbed ('By the mark ten').

'Stabilised,' confirms the Captain. The auto-throttles have captured the final approach speed – 163

knots and Concorde is established in the correct position for landing.

'Four hundred feet,'

'One hundred to go,' responds the Captain

'Three hundred feet,'

'Decision height,' calls the Captain.

Landing clearance has been received, the runway is visible and clear of obstacles. 'Continuing,' responds the First Officer. A 'go-around' is possible right up to the point of touchdown.

The approach lights on runway four right protrude out of the water of Jamaica Bay. 'Two hundred feet,' and the pilots are 37ft higher than the main wheels. 'One hundred feet,' over dry land now. 'Fifty,' the auto-throttles are disconnected. 'Forty, thirty, twenty, fifteen.' The throttles are manually closed. At this point the aircraft would be pitched down, both by the pressure of air trapped between it and the runway, referred to as 'ground effect', and by the reduction in thrust, but the First Officer gently checks this tendency by bringing the control column back to hold the attitude constant as the descent rate decays in the increasing ground effect. He also pushes the left rudder pedal to lose the remaining 3 degrees of drift required to fly down the centre of the runway, caused by the crosswind. Now the main wheels track along the runway for a moment before the touch.

A puff of smoke and dust whisks from each set of main-wheel tyres into the two vortices, one over each wing: a clear indication of the nature of the airflow that has been supplying lift during the final stages of the approach. Reverse thrust is selected whilst the next manoeuvre is carried out – landing the nose wheel; during this manoeuvre the flight deck has to descend a further 17ft.

Top: View of runway 4R at JFK from about 1,000ft on final approach over the lowered nose and visor.

Middle: At 100ft.

Bottom: Crossing the runway threshold. The wheels are still at 50ft above the ground with the pilots another 37ft above them.

'Stick forward,' calls the First Officer. The nose wheel is down, and reverse thrust is selected. There is a roar in the cabin – noisier than at take-off. The brakes are applied. The stopping appears urgent, but all is normal. 'One hundred knots,' calls the Captain. Outboard engines are selected to idle reverse. 'Seventy five knots,' and the inboards go to idle reverse. 'Forty knots groundspeed,' all engines are selected to forward idle thrust.

'After landing check, shut down two and three,' says the First Officer. The nose is selected to 5 degrees, unnecessary systems are switched off, the two inboard engines are shut down and the Engineer pumps four tonnes of fuel into the front tank to ensure that the aircraft will not tip tail down during unloading.

'Left at the end, cross one three left, right on the outer and call ground point nine,' says the controller. It is a kind of pidgin English, meaningless to the uninitiated, but totally clear to the crew. Over the maze of concrete and tarmac that constitutes John F Kennedy Airport, New York, the crew find their way to the BA Terminal, at a speed no faster than a mile every two and three quarter minutes.

'Ladies and gentlemen, welcome to New York.' This time it is the Captain addressing the passengers.

'Our maximum altitude today was 58,000ft and our maximum speed 1,320mph, giving an average over the whole distance of close on 1,100mph. On behalf of all the crew thank you for flying with us on Concorde; we look forward to seeing you all again. Finally, the jumbo jet that left 10 minutes before us from Heathrow is very nearly, but not quite, half way here.'

A marshaller waving two fluorescent wands directs Concorde over the last few feet. The wands cross, the aircraft stops.

'Parking checklist,' says the First Officer.

'Brakes,' responds the Engineer.

'To park,' replies the First Officer.

The checklist continues, like some litany. The passengers disembark. 'Mind your head sir as you leave,' cautions the Cabin Services Director, standing at the forward door to bid 'au revoir' to the passengers. They have had an excellent flight.

Opposite: The landing sequence. Concorde G-BOAF landing on runway 4R at Kennedy. Note the vortices on touchdown, the landing of the nose wheel and the 'buckets' (secondary nozzles) closed over the jet pipes to deflect the jet efflux forward at an angle of about 50 degrees to the horizontal to give reverse thrust. Since the aircraft tends to pitch up with reverse thrust selected the stick is held forward, thus the elevons move down, as can be seen. The nose is in its fully down 12½ degrees position for the landing, once the landing run is complete it is selected to 5 degrees. Finally Concorde taxis into the BA Terminal at New York JFK.

TRAINING OF THE CONCORDE PILOT

BA, like most of the world's airlines, only employs pilots who are already qualified. They must have at least a Commercial Pilot's Licence (CPL), be rated to fly a twin-engined aircraft and have an Instrument Rating. The latter allows a pilot to fly an appropriately equipped aircraft in cloud and into certain types of 'controlled' airspace, whether or not he can rely on visual reference. A private individual can pay for his (or her) own training at a Civil Aviation Authority 'approved' flying college – a very expensive procedure. Or he can become qualified by acquiring 700 flying hours (200 hours on the 'approved' course) and passing the relevant exams. Although cheaper, the 700-hour route to a professional licence can take several years.

If he (which includes 'she' in this chapter) is very lucky, he may find that airlines in general, and BA in particular, are recruiting cadet pilots. If this is the case he can apply for BA sponsorship at one of the CAA 'approved' flying colleges. For this there are certain criteria: typically he must be between 18 and 24 years old at the start of training, be medically fit and be predicted to stay that way during his career (up to 55 in BA, but 60 is the UK upper age limit for most commercial flying). Furthermore, potential pilots must have the necessary exam qualifications and pass the interviews, along with coordination and aptitude tests. Usually, the exam qualifications include two 'A' levels, preferably physics and maths, but he must have passed at least maths and a science subject, along with English

and two others at GCSE. A university degree may waive some of the specific requirements described above.

I had been lucky enough to receive sponsorship from BOAC for my training at the College of Air Training Hamble. In those days we could choose which of the two nationalised airlines we would like to fly for, BEA (British European Airways) for European flights or BOAC for worldwide flying. Clearly BEA was not going to order Concorde, accordingly I chose BOAC. After an 18-month course at Hamble I found myself starting at BOAC's training school at Heathrow on

The external world in this picture is of Kai Tak, the former Hong Kong airport. The image here is generated by a computer and extends to 165 degrees around the cockpit, which replaced the original system. Although much more realistic especially when generating a night or dusk view, it was still not sufficiently realistic to allow the student to perform his first landing on a scheduled service. There still was a requirement for the converting pilot to fly no less than 14 circuits with 'touch and goes' – in other words not stopping having landed.

Above: Proof of motivation is essential for any pilot seeking training sponsorship from an airline. However, basic training does not include hang-gliding. Delta-wing gliders of the 1920s were developed to reduce drag through absence of a tail plane. On these little fabric winged deltas at the Grand Salève near Geneva, the upturned wing tips restore the stability normally provided by a tailplane. The effectiveness of swept wings for flight as speeds approaching Mach 1 was recognised by the Germans during the Second World War and only after the war by the Allies; not that the 'deltiste' (hang glider pilot) has any intention of flying faster than sound.

Left: The members of the flight deck crew were equipped with 'quick-don' oxygen masks capable of supplying 100% oxygen under pressure. The donning of oxygen masks and the emergency descent was practised regularly in the flight simulator. If the loss of pressure above 50,000ft were very rapid, the crew would 'pressure breathe'. Instead of breathing in, a conscious effort to breath out was required. This technique ensured that the members of the crew did not lose consciousness. Speaking to each other over the intercom sounded rather odd and needed to be practiced annually.

3 March 1969. On the previous day I had watched on television the maiden flight of Concorde 001 at Toulouse, *quelle joie*, as they say in France.

Once in the airline, the BA pilot can choose, within certain constraints such as seniority, the type of aircraft that he wants to fly. If that aircraft should have been Concorde certain extra rules applied. He would have already been qualified on one of BA's aircraft, which meant in practice he would have been in the airline for at least three years. He would, in order to justify the extra training expense, have been able to have served sufficient years before promotion or retirement, usually seven.

There are four main parts to any conversion course: 1) the technical course, plus written examinations; 2) the simulator course, plus compulsory tests; 3) 'base'

training – flying the aircraft without passengers – also with tests; and 4) flying under supervision with passengers, plus the final 'route' check. The pilot then becomes fully qualified to fly as captain or co-pilot.

During the six-week Concorde technical course for pilots (eight for Flight Engineers), each system – hydraulics, electrics, flying controls and undercarriage – had to be thoroughly learnt. On the B747-400 this course lasted a mere three weeks. The information for most aircraft conversion courses in BA since the 1970s is presented through means of an 'audio visual' – slides accompanied by audio tapes. On Concorde the 'audio' part was taken by a lecturer, which allowed dialogue that could highlight, for instance, how a system evolved – a helpful element in the learning process. During the Concorde technical course the simulator was used as a

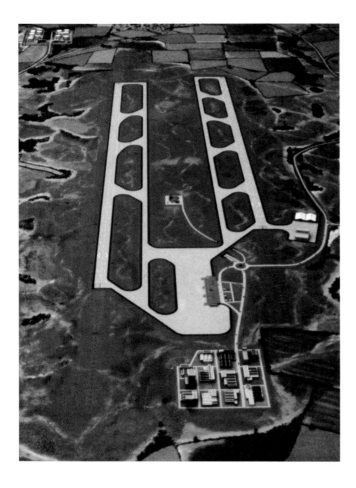

When first commissioned at Filton the Concorde simulator had a visual system which consisted of a television camera flying over a model airfield. The picture seen by the trainee pilot was collimated to infinity, but was in 'narrow vision'. It was not realistic enough for teaching the trainee how to land. Curiously in 1976 when first commissioned, the Concorde simulator cost as much as a VC10 airliner to buy – about £3½ million and the same per hour to operate – c. £400.

'systems trainer', illustrating theory with simulated practice. Finally, the CAA-administered exam took the form of a multiple-choice test on each system.

On Concorde the simulator training lasted 76 hours – 19 four-hour details (50 hours are spent on the B747-400). Every emergency procedure was practised until perfect. Single or double engine failure, pressurisation failure followed by emergency descent and failure of electrical signalling to the flying controls, are examples of some of the procedures. The trainee had to pass several tests on the simulator. Finally, he had to renew his Instrument Rating – the general

qualification which is not normally specific to a particular aircraft type.

The visual and motion systems on some aircraft simulators qualify the converting pilot to demonstrate his first landing on that aircraft for real with passengers on board – albeit under the supervision of a highly experienced training Captain. Originally the Concorde simulator (then at Filton, Bristol) made use of a television camera 'flying' (suspended) over a model runway and environs within circuit distance. The image captured was projected onto a screen in front of the

Opposite: Concorde on final approach. The trainee Concorde pilot must fly at least 14 circuits before his/her route training. At high angles of attack – 14 degrees on final approach to landing – drag increases (demanding extra thrust) with reducing speed. Experiencing both the 'feel' of the actual aircraft, and the look of the runway in reality, arms the pilot with the confidence needed to tackle his/her first landing with passengers. (Mark Wagner/aviation-images.com)

Training on the simulator of the two-crew Boeing 747-400. Concorde, in spite of being more complex, was developed just before information presented in cockpits via television screens became commonplace. Outside, the computer-generated image is stunningly realistic.
(© Frederic Pitchal/Sygma/Corbis)

cockpit, but only with a limited horizontal field of vision, albeit collimated to infinity. By 1989 the original system had been replaced by a 165 degree field of view, colour, daylight, dusk or night computer-generated image. It was capable of showing individual airports – like Kai Tak Hong Kong with its famous curved approach path, or London Heathrow complete with gasometer to the north-east as well as the airport buildings – to give realism during taxiing. In addition, the Concorde trainee had to carry out about 15 landings during 'base' training. On the B747-400, only if the pilot's previous type has been significantly different from the Boeing, must he land the real aircraft, usually a minimum of four times.

Take-off in the real Concorde (like the simulator), was commenced with the call, 'Three, two, one, now' when the throttles were moved smoothly and swiftly from their idle position to fully open. The four columns of engine gauges reacted with whirling digits and a single synchronised swing of 20 needles. Next, the four primary nozzle area gauges moved into the white segment, proof that each reheat had lit. The brightness of the outside and the force of the acceleration were evidence that this was no simulator. Each Concorde pilot flew three 'details' totalling 2¾ hours (30 minutes at night) and carried out at least 14 circuits. Most were 'touch and goes' (landing without stopping followed by immediate take-off), while a few had to be 'go arounds' (approach to less than 200ft above the runway followed by a climb out). On about half a dozen occasions there was a simulated loss of thrust from one engine on take-off. Once all

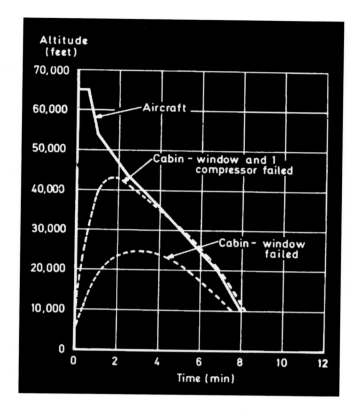

Above: A stream of particles emitted from the sun in a solar flare might have exposed the occupants of Concorde to a radiation hazard. Should the rate have exceeded 50 millirems per hour (the red quadrant) a descent would have been called for, levelling off once the warning had ceased but no lower than 47,000ft. There was an unconfirmed report that a small amount of radiation following the Three Mile Island nuclear power station accident of 1979 was detected by Concorde.

Left: Should there be a problem with the pressurisation, the aircraft would carry out an emergency descent (continuous white line). Should a window fail, the cabin altitude would rise as indicated by the lower broken white line – unless an air conditioning group had been shut down.

the items were satisfactorily completed, the pilot went on to the final stage – route flying.

Now the trainee, under supervision and with an extra crew member to 'cover' for him, operated his first scheduled supersonic sectors. To begin with on the flight deck there hardly seemed to be enough time to fit in all the pre-departure tasks: air traffic control clearance, take-off calculations, check lists, briefings, loadsheet, technical log, the problems caused by a late passenger, a minor technical defect – was there time to correct it, or did it limit performance? Relief at last, the door closed and it was time to depart. At least the flying was not so different from that received during training – confidence returned.

As the sectors clocked up, so the tasks seemed to require less time. Finally at the end of 20 sectors for a Captain and 16 for a First Officer the 'route' check was carried out. Once successfully over this last hurdle, the pilot suffered no more formal supervision. He was now

qualified until the next six-month check on the simulator and the next annual 'route' check on the aircraft. On average six months elapsed from the start to finish of a Concorde course, in comparison to about three-and-a-half months for the B747-400.

In April 1993 the first lady pilot ever to be qualified on Concorde took to the supersonic routes. Barbara Harmer, a former British Caledonia pilot, became a Concorde First Officer. By June 1996, 108 BA pilots and 41 flight engineers had successfully completed the Concorde conversion course. The final total in October 2003 was 134 BA pilots and 57 flight engineers (see Appendices).

IN SERVICE

In 1903, man achieved the 'impossible' – flight in a machine heavier than air. Then, in spite of enthusiastic conjecture to the contrary, most serious commentators foresaw neither supersonic nor space flight. Yet only 66 years after the Wright Brothers' feat at Kitty Hawk, a supersonic airliner had flown and man had landed on the moon. Next it seemed as if all long-range flight would be supersonic and that the moon would be colonised by the turn of the century. However, in 1976, when Concorde entered into service, its prospects appeared gloomy. By 1986 improvements were apparent. This chapter was first written after Concorde had been in service for nearly 21 years, by which time there had been a steady growth in subsonic passenger numbers.

This passenger growth has stretched resources. Reformed administration benefited by computers has enhanced air traffic control and airports have expanded. Nevertheless Heathrow was, by 1996, in urgent need of the proposed fifth terminal. In the year 1995–1996 54 million passengers were in transit through Heathrow, almost 30 million of them with BA.

By 2011, with Terminal 5 now in active use, the figure is 80 million with BA maintaining its share. Without this expansion Heathrow, in spite of having celebrated 50 years of pre-eminence, would have lost business to a rival continental airport.

Heathrow's 50th birthday was celebrated with a fly-past of historic and present day aircraft. For half an hour during the afternoon of 2 June 1996, the business of transporting passengers was halted. Leading the assortment of aircraft was the Avro Lancaster, followed, amongst others, by two diminutive de Havilland Dragon Rapides; the elegant Comet, representing the first of the jet airliners; Concorde, in formation with BAe Hawks of the Red Arrows and finally the then most recent subsonic jet airliner, the massive twin-engined Boeing 777 of BA.

After nearly 21 years Concorde still evoked wonder. For the celebration described above, the BA flagship was under the command of Captain Michael Bannister, the Flight Technical Manager, but was flown from the left-hand seat by Captain David Ross, himself a former RAF pilot. As such displays require exact navigation and precise timing, especially the link up with the Red Arrows (the first for ten years) the operation was closely monitored by Senior First Officer Tim Orchard, whilst the flight engineer's panel was operated by Ian Smith – a long-serving Senior Flight Engineer on Concorde.

The BA Concorde operation accounted for a small proportion of the 150,000 passengers that then passed daily through Heathrow. But Concorde carried a significant group of people. It has been noted by some observers that the performance of the UK economy throughout the 1980s and 1990s could almost be predicted by an analysis of Concorde's revenue.

Concorde's principal scheduled services were from London, Paris or New York. Over the winter months there was, for BA, a once, sometimes twice, weekly scheduled service to Barbados. In 1980 BA, in association with Singapore Airlines, discontinued the London/Bahrain/Singapore route. In 1982 Air France ceased Concorde flights to Rio de Janeiro, Caracas, Mexico City and Washington. In October 1994, after 18 years, BA dropped their London to Washington service, the Miami extension to that route having been discontinued by January 1991. By the cutting of the latter two routes, BA saved 600 supersonic sectors per year thus prolonging Concorde's useful years of service.

With the cessation of the direct thrice weekly Concorde service to Washington Dulles, a daily morning connecting flight was established between New York and Washington National. This took about 90 minutes longer than the old direct flight. In compensation 'National' was 45 minutes closer by road to the US capital than 'Dulles'.

In good economic years Concorde was extremely profitable. It even survived the recessions of the mid-1980s and early 1990s. In 1996 BA's twice daily service between London and New York, at normal Concorde fare levels, was achieving load factors westbound of 70–80% and eastbound 50–60%. (In June 1996 the return fare was £5,774 or £4,772 with $70 cancellation charge, versus £4,314 subsonic First Class; the cheapest Concorde charter ticket being £225 for a 30-minute subsonic and £545 for a 100-minute supersonic flight.) For Air France the single daily service between Paris and New York achieved an overall load factor of over 60%. By October 1987 the BA Concorde had carried its 1 millionth scheduled transatlantic passenger; the figure for Air France being 677,000. By the 20th anniversary of Concorde operations (21 January 1996), Air France had carried a total, including charter passengers, of over 1.1 million; for BA this total approached 2 million.

The Red Arrows formate on Concorde as they approach Heathrow for the airport's 50th anniversary flypast. Concorde was commanded on this flight by Captain Mike Bannister. (John Dibbs/aviation-images.com)

This publicity photograph of four Concordes flying in formation was taken on Christmas Eve 1985. Leading the formation in 'AA' was Concorde general managaer Captain Brian Walpole, with Captain John Eams commanding 'AC', John Cook 'AF', and David Leney 'AG'. (Airbus courtesy BA)

Concorde carried more passengers west than eastbound. The supersonic traveller from Europe arrived in New York one hour, from UK, or two hours, from France, before he had left. Westbound the time change was subtracted from the flight time; eastbound it was added. It would in theory be possible to start a round the world supersonic westbound service with a single aircraft that always left at the same local time.

The reason for the disparity is as follows. Having arrived in New York on the morning supersonic flight, many businessmen completed a full day's work. Then despite the jet lag, they returned on the overnight subsonic flight, to be ready (or so they thought) for work the next European morning.

A less tiring option, although involving an overnight stay, was to return on the following morning or afternoon supersonic New York to Europe flight. Leaving at 8.45am (local time), on flight BA002 the executive returned to London at 5.25pm (BA004: 1.45pm–10.25pm; Air France New York–Paris on AF001: 8.00am–5.45pm).

For the passenger who originated his trip in the United States, the supersonic day flight was very attractive. This delivered him fresh in Europe allowing easy adjustment to the new time zone.

Concorde was constrained by 'night jet bans'. At Heathrow no aircraft 'movements' were allowed between 11.30pm and 7.00am, at New York between 10.00pm and 7.00am. Theoretically it was possible to avoid the ban at each end, nevertheless an overnight supersonic eastbound flight was impracticable. A 9.45pm take-off from New York (leaving very little margin for error) could often arrive at Heathrow before 7.00am. Reducing speed to arrive at, say, 7.30am, would have defeated the object of Concorde, furthermore there could still have been complaints about early morning jet noise. In any case, having suffered a very short night's sleep and paid more for the privilege, the passenger could be less comfortable than his subsonic counterpart. For a Concorde successor flying eastbound, there was a potential conflict of night jet bans with the short flight times. This conflict lessened as the local hour change between points increased. An eight-hour time change between two places with a five-hour flight, would make for an eastbound journey equivalent to 13 hours. So a flight departing at 8.00pm would arrive at 9.00am the following day, both times being well within jet ban tolerances.

When Concorde services started, a cartoon was published showing a very jet-lagged top executive being consoled by his secretary. She was saying to him: 'Look at it this way, you may have lost the multi-million dollar deal, but you did save one hundred and sixty-two pounds and twenty pence by not taking Concorde.'

In BA charter flights generated 10% of the total revenue from Concorde. In 1996 seven companies

regularly chartered Concorde: Cunard, Goodwood Travel, Concorde Spirit Tours, Superlative, Intrav, David Gladwin Concorde Ltd and Yorkshire Charters.

Cunard offered an Atlantic crossing by QE2 one way and Concorde the other. The QE2 passengers were frequently entertained by a Concorde crew member who naturally extolled the virtue of the two extremes of speed.

Goodwood were formed in 1981 by Jan Knott, Colin Mitchell and George Stevens and are based in Canterbury. Their first 'Flight of Fantasy' took passengers to Nice, France for the 1983 Monaco Grand Prix. Concorde Spirit Tours, founded by American Attorney at Law Donald Pevsner, broke the circumnavigation of the globe record. On 12–13 October 1999 Air France Concorde F-BTSD completed the trip around the world in 32 hours 49 minutes and three seconds. In June 1994 Goodwood in conjunction with Cunard's QE2 combined two anniversaries: the 25th of Concorde's maiden flight and the 50th of the D-Day landings on Normandy. By August 1995 Goodwood had carried 70,000 passengers on Concorde. By June 1996 they had taken Concorde to no fewer than 47 destinations (see Appendices for all Concorde destinations). In particular Goodwood were famous for the 'Father Christmas' Concorde flight to Rovaniemi, Finland. They were noted for their superb brochure and the imaginative way they rekindled the romance of travel. They offered such combinations as:

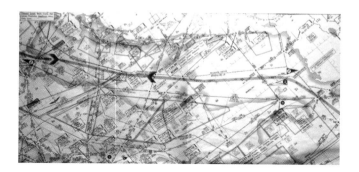

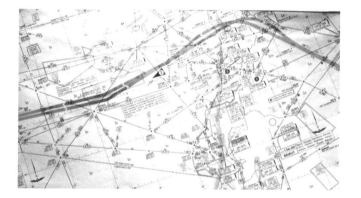

Left: On the London/Bahrain sector, the first chart shows the supersonic track over the Adriatic. It is carefully routed to avoid placing a sonic boom on either Italy or Albania, a minimum of 30 nautical miles (nms) from the coastline is required on the inside of a turn and 20nms flying straight. The track fits precisely within the limits. During the troubles in former Yugoslavia, the route was temporarily suspended. The second chart shows the eastbound and westbound tracks to the south of Cyprus and over the coast north of Beirut. Until 1980 supersonic flight was permitted over North Lebanon. Subsequently Egypt permitted supersonic flight over the Nile to the south of Cairo; the new route then crossed Saudi Arabia to Bahrain. Later, approval of supersonic flight over Saudi was withdrawn, another factor in the decision to discontinue the Singapore service which had operated, off and on, for barely two years.

Concorde to Istanbul with a return on the Orient Express; Concorde in conjunction with the Golden Arrow; and Concorde to Jordan with an excursion on the train used by Lawrence of Arabia. Also on that tour was a visit to Petra – the Rose Red City – which was carved out of the living rock. 20 centuries later a similar technique was used in the construction of Concorde – but instead of rock the medium was the carefully developed Hiduminium RR58 aluminium alloy, out of which were carved pieces of fuselage by integral machining.

Superlative Travel, based in London, also featured the exotic factor of Concorde travel to romantic destinations – Voyage to Monte Carlo, Mediterranean Magic and Caribbean Christmas. Both Superlative and Goodwood Travel liaised with Cunard's QE2.

Intrav was based in St Louis, Missouri. Their speciality was in organising round the world Concorde charters at the rate of two or three per year.

David Gladwin Concorde Limited was based in Nottingham. A former BA VC10 pilot, Gladwin was scheduled for a Concorde conversion course in 1978. Difficulties within BA caused the course to be cancelled. In 1986, instead of becoming a Concorde pilot, he started chartering Concorde. Ten years later his firm had carried almost 40,000 passengers. Included have been special flights, one to inaugurate Manchester's new terminal and others marking the 25th anniversary of Concorde from Filton, Bristol – Concorde's birthplace.

The smaller end of the Concorde charter business was, in 1996, being conducted by an enthusiastic company – Yorkshire Charters managed by Keith Walker and based in Ossett. In April 1987, they arranged the first visit of Concorde to Leeds Bradford. This proved so successful that it was repeated several

The three icons, Concorde G-BOAG, the Red Arrows and the QE2 over the English Channel, summer 1985. Captain John Hutchinson, noted for BBC television commentary for major British air shows, was on board on this occasion, with Captain Leney (Flight Manager, Technical) at the controls. Very many Concorde charters used to be organised by Cunard – hence the publicity value of this association.

The Crown Area Modification called for an extra strap over the top of the fuselage and the removal of a double row of fasteners (in two positions over the fuselage) for strengthening through 'cold working' prior to the refitting of the fasteners.

The floor looking aft in the vicinity of the forward galley. Even the most minute blemishes caused by corrosion were replaced. The section on the right with the circled area, where the corrosion has been removed, is to be substituted with the one on the left which has been newly assembled.

times in subsequent years in the 1990s, with Newcastle and Teesside being included on other occasions.

How did Concorde perform technically? Brian Calvert, in his book *Flying Concorde*, likened the building of Concorde to watch making in a hangar. The servicing of the 'time machine' was somewhat similar.

At specific intervals, Concorde underwent a series of checks, but every 12,000 flying hours each aircraft was submitted for a 'major' check. In BA the last aircraft in the fleet to achieve 12,000 hours was G-BOAG (aircraft 214). By 2004 the series of 'majors' was due but this was preceded by the retirement of the entire fleet. A 'major' on Concorde lasted about five months, which included a month for the structural work associated with the Crown Area Modification.

The Crown Area Modification strengthened the top of the fuselage. At 'rotate', during take-off, the elevons moved up placing a down force at the tail. The reaction to this stretched the top of the fuselage as the aircraft was pitched up. Calculations suggested the need for minor reinforcements along this section. In spite of this, a computer model was made of the structure of 'G-BOAF'. About 40 strain sensors were fitted and the data they gathered over several flights was analysed. A far more accurate picture of the structure was formed

than was possible when Concorde entered service. This proved that the designers had erred on the right side. Concorde was exceedingly strong, albeit at the cost of a small weight penalty.

Before the Crown Area Modification could begin, the aircraft had to be 'candlesticked'. This involved being jacked up in such a way that the skin experienced zero stress; adjustments were even made for diurnal temperature variation. Then at two positions over the top circumference of the fuselage a double row of fasteners was removed. Into each resulting hole was forced a mandril. The ensuing stress strengthened the surrounding metal. The mandril was then removed and a new fastener put in place. Since all this took place at room (or hangar) temperature, it was called 'cold working'. Included in the modification was the addition of one strap over the top surface of the fuselage just to the rear of the centre door.

With the intention of keeping Concorde flying well into the future, the Life Extension Programme was also introduced. Bitter had been the experience of the builders of the early Comets – at least two of which, in the early 1950s, had suffered explosive decompression. Hence the requirement, already mentioned in the chapter 'The Turn of the Tide', to build two Concorde fatigue specimens and subject them to forces, vibrations, heating and cooling to an excess of that expected during normal service. Before being shut down in 1985, the test rig at Farnborough achieved 20,000 'supersonic flight cycles'. After 6,700 cycles – a third of the total – it was agreed that more stringent airframe inspection procedures would have to be established. These constituted the Life Extension Programme.

Rather than the supersonic flight cycle, the yardstick established for measuring Concorde life subsequently became the reference flight (RF). Every flight was counted but the figure was modified in proportion to the take-off weight. A flight at over 170 tonnes take-off weight counted as one RF and less than 120 tonnes as half an RF. By 14 June 1996 the oldest BA Concorde had achieved 6,042 RFs and the youngest 3,792 RFs. The initial Life Extension allowed for 8,500 RFs. Three years before this total was reached, a review would take place, modifications made (if necessary) and different inspection procedures (if required) established to take the life up to 10,000 RFs. The oldest Concorde was due

to achieve 8,500 RFs by the year 2004 and the youngest by 2012. With an extension to 10,000 RFs and assuming no other limitation, Concorde could have still been flying up to 2020. A minimum of four Concordes would have been needed to support the New York service.

To make the various inspections required by the Life Extension Programme simple, modified aircraft were fitted with borescopes. These are visual devices – using fibre optics or even micro TV – capable of seeing round corners to examine the more inaccessible parts of the fuselage without dismantling internal structures such as galleys.

Due to thermal heating during the supersonic cruise, Concorde's fuselage was less prone to corrosion than that of a subsonic aircraft. Fluids evaporate from the inside of an aircraft skin regularly heated to 100°C (212°F); in contrast they condense on the inside at −30°C (−22°F). On Concorde it was areas where fluids

A Concorde 'First Day Cover', this one celebrates the start of supersonic commercial services to America.

HM The Queen Mother's 85th birthday present from BA was a trip in Concorde on 6 August 1985. Also seen here are Captain Walpole (General Manager, Concorde Division 1982–1988) and Senior Engineer Officer Peter Phillips.

could be spilt onto the structure at room temperature, for instance around the galleys, that needed to be subject to greater scrutiny. During checks very minor blemishes were 'blended out' by careful metal polishing; if larger, then the piece was completely renewed.

There was also the question of whether there would always be a sufficient supply of spare parts over the projected lifetime of the aircraft. When necessary new parts had to be built from scratch. For instance new rubber sheathing to the weather radar cable guides had to be made out of a non-carcinogenic material and electronic circuit boards for the air intake computer units were constructed as new items.

When the rudder surfaces on the fin showed a tendency to delaminate, new sets had to be manufactured.

The £7 million bill included the construction of new manufacturing jigs as the original jigs had, in 1981, been scrapped. The new jigs were kept with Concorde 202 in a hangar at Filton.

Just after Concorde went into service three important modifications were made: thinning of the lower intake lip; approval of a further aft centre of gravity for take-off; and stiffening of the main undercarriage oleo legs. The first modification improved supersonic range, enabling the Bahrain/Singapore sector, at the time routed to the south of Sri Lanka, to be flown direct. The second allowed more fuel to be carried in the rear fuel tank thus improving take-off performance; this is due to the increased 'flap effect' of

Opposite: The Queen's Golden Jubilee saw a formation flypast by the BA Concorde G-BOAD and the Hawks of the Red Arrows, trailing red, white and blue smoke over the Houses of Parliament and Buckingham Palace. (Mark Wagner/aviation-images.com)

On 2 April 1996, Pepsi-Cola launched their new 'livery'. Air France Concorde 'F-BTSD' in the new Pepsi colours visited ten European and Middle Eastern cities, including here Dublin. Celebrities Claudia Schieffer, André Agassi and Cindy Crawford assisted at the launch held at London Gatwick Airport. Contrary to press reports, the paint withstood supersonic flight. The aircraft was restored to Air France colours after two weeks. (Jonathan McConnell)

the elevons (angled further down to compensate for the more rearward centre of gravity). The third, a modification to the undercarriage, allowed Concorde to use some of the more undulating runways around the world. During the route proving phase (prior to entry into service), violent oscillations had been experienced on take-off at Singapore.

A later modification involved fitting strain sensors to the main undercarriage bogie beams. Should a tyre have deflated below a certain speed during take-off, the twist on the beam would have caused a warning light to illuminate calling for the take-off to be abandoned. This avoided overstressing the neighbouring tyre, risking a burst and debris entering the engines.

Apart from an improvement in the computer software of the Inertial Navigation System and the fitting of the Traffic Alert and Collision Avoidance System (TCAS) no other significant modifications were made.

Occasionally changing legislation means that airliners must be fitted with new avionics. After a specified date, the United States would not accept an airliner, even Concorde, into its airspace without TCAS. First somewhere had to be found on Concorde's densely packed instrument panel for the display; then the detector aerials had to be fitted. A solution was found for the former; but due to becoming overheated during supersonic cruise, the aerials refused to work. The deadline passed… Months of special exemption were in fact allowed before the problem was finally solved.

Could the electromechanical instruments (artificial horizon, compass etc.) on Concorde have been replaced with television screens? These are found on later airliners with computer generated display images of similar but improved instruments. This was possible; but very expensive. The expense lay not so much in the purchase and fitting of the equipment; but in the necessary certification process – six months of static testing, six months of test flying without passengers followed by one to two years of revenue flying.

BA used all seven of its Concordes, while Air France, which originally had seven, used five out of its six on a rotational basis. In November 1977, F-BVFD (211) was involved in a heavy landing at Dakar – 14 feet per second (fps) at touchdown (10 fps

is the standard limit). Only minor repairs were required, further evidence of the immense strength of the structure. In 1982 Air France finally grounded 'FD'. Out of service for 12 years the aircraft had suffered serious corrosion and was dismantled in 1994.

Improvements were made in flight planning. Once determined manually, the fuel calculation was later made by computer. This more accurately accounted for the forecast winds and temperatures of the day. The Point of No Return (PNR) chart was similarly treated. Concorde would have lost 25% of range if some failure, such as an engine shut down, forced a deceleration to

On 2 March 1989, an invited audience at Toulouse celebrated the 20th anniversary of the first Concorde flight by 'son et lumière'. Note the musicians in an acoustic glass box by the nose wheel of the psychedelically lit F-WTSB (Concorde 201). Although the mood was generally upbeat, the message was that there would be no immediate successor to Concorde. (Airbus)

subsonic speeds. Unable to make the destination or return to the departure airport, the chart showed the alternates that can be safely reached from various points across the Atlantic.

The routes to and from the United States have, over the years, remained essentially the same and are shown on the endpapers of this book. As data was accumulated

about the propagation of the secondary boom and its variance with the strength of the upper winds, deceleration points were increasingly subject to seasonal modification.

When Concorde entered service, critics were quick to point out that the equivalent of half the flight time was 'wasted' in airport procedures. Although there

were many subsonic flights of similar duration to Concorde's, supersonic flight seemed to emphasise the incongruity. With this in mind, the BA Concorde Brand Management Team arranged everything to cut the red tape. There was a dedicated fast track channel, a telephone check-in for those who carried only hand luggage and a suiter service. This ensured the suit

carrier reached the baggage hall within eight minutes of arrival. Finally there was the famous Concorde lounge which was remote from the ceaseless commotion of the airport.

On board there were six cabin crew members, all of whom had been especially selected for their suitability and dedication to operating Concorde. All 236 Concorde qualified personnel also served on BA's narrow body fleet aircraft (Boeing's 737 and 757 and the Airbus A320), 60% were seconded for two years, the remainder for longer periods. This mix achieves enthusiasm tempered with experience.

The 'technical crew' consisted of two pilots and a flight engineer. In 1996, to fulfil all the planned flying, about two and a half crews per aircraft were needed. The list in the Appendices shows more Captains 'current' than First Officers. This was due to a retirement bulge having (in June 1996) called for an increase of Captains to cover for the imminent retirement of some of their colleagues. On average a flight deck crew flew eight trips (of two days) per month, had standby duties, ground training and also public relations duty days. The allocation of scheduled flights was decided on a seniority basis, while the charter flights were shared out. Six days per year were spent on the simulator for checks and refresher training.

Concorde's role continued to evolve. The scheduled routes of both its airlines were gradually pared down

Concorde at Waco in Texas during a charter visit between 12 and 15 June 1986, organised by Columbus World Wide Travel. James Hamilton (later Sir James), as Director General of the Concorde Division at the Ministry of Aviation from 1966 to 1970, feared that political pressure might force Concorde into service before being thoroughly tested. Such had been the fate, he noted, of the British Airship R 101 which crashed on 5 October 1930 at Beauvais in France on her maiden flight from Cardington (England) to India. Among those killed was Lord Thomson of Cardington, Secretary of State for Air, and possibly the next Viceroy of India. Mercifully the shadow of the past, in this case of the Goodyear Airship, did not come to haunt the Concorde project. Nevertheless, sales might have been far greater if the original in service year of 1969 had been achieved rather than 1976.

Six Concordes at Heathrow, Boxing Day 1986. Over the Christmas period it was possible to muster almost the entire fleet in one place. (Airbus courtesy BA)

and were, by 1996, at a minimum consistent with profitability. The charter market, for which Concorde was not originally designed, but in which it had been notably successful, remained buoyant. Before the end of its life Concorde would be flown by pilots born long after it entered service.

Although older in years than most of the rest of the BA fleet, Concorde was not old in terms of utilisation. In 1996 the average Concorde had flown the same number of hours as a four-year-old Boeing 747, achieved the same number of flights as a two-and-a-half year old Boeing 737 and was in the same condition as a three-year-old Boeing 757. That Concorde was kept in such superb condition was due, in particular, to the dedication and expertise of the ground engineers of both BA and Air France.

TRAGEDY AT GONESSE

It was Tuesday of Air Show Week at Farnborough. I had left the show early, and was walking to the station to catch a train to London for a reception at the Royal Aeronautical Society. My daughter, who happened to be watching BBC News 24 at the time, phoned me to say that a Concorde had crashed. With only 12 examples flying, it was an accident that statistically was never supposed to happen. Were there any survivors? Where did it happen? Which airline was it? Back came the replies: Possibly, Paris, Air France.

On 25 July 2000 at 14:43 GMT, the tyre of the number 2 wheel of Concorde F-BTSC burst during take-off. This set in motion a train of events that resulted in the destruction of Concorde 203. Two minutes after starting the take-off roll, the aircraft crashed on to a hotel at Gonesse, 8km to the west of Charles de Gaulle airport, with the tragic loss of all 109 people on board, four on the ground and injury to six others.

Early reports had spoken of survivors; these turned out to be hotel staff and guests. One girl had had a miraculous escape. The owner of the burnt-out hotel spoke of the extraordinary sound that Concorde had been making: '…really overdoing it.' Other observers referred to the bravery of the Captain who turned his burning aircraft away from a more densely populated area. *The Times* reported how the crash was witnessed by a group of 65 young musicians from Suffolk travelling to the hotel.

News of the accident reverberated around the world. Journalists gave instant analyses, while Jim Naughtie of BBC Radio 4's 'Today' programme was sent to Paris. At Farnborough the flags were lowered to half-mast.

BA cancelled that evening's service to New York and, after some agonising, resumed Concorde services the next day whereas Air France did not.

In the days that followed, details were revealed about those on board. Of the 100 passengers on Special Flight AF4590, one was an American, one an Austrian, two were Danish and 96 were German. All were en route to New York where they were to join their cruise ship, the MS *Deutschland*.

On Thursday 27 July, a memorial service was held at L'Eglise Madeleine in Paris, the same church where, three years earlier, people had paid their last respects to Diana, Princess of Wales.

At first the Civil Aviation Authority did not withdraw Concorde's Certificate of Airworthiness. However after pressure from the equivalent French authority, the Direction Générale de l'Aviation Civile (DGAC), Concorde was grounded by both countries.

It should be noted that the design of Concorde was subject to the airworthiness requirements that were drawn up in a document entitled the Transport Supersonique Standards (TSS). Because of Concorde's revolutionary aerodynamics and speed range, these requirements were often more stringent than those for a subsonic aircraft.

With respect to an engine failure during take-off TSS required that the climb gradient should be 4% or more whilst climbing at the engine out safety speed – V_2; in comparison to 3% for a subsonic aircraft at its V_2. The reason being that the performance of the slender delta deteriorates more sharply when an incorrect speed is flown compared to a conventional swept wing.

The maximum all up weight on take-off is limited by:

1   Maximum structural weight, just over 185,000kg for Concorde.
2   Performance. The limiting case being when at decision speed – V_1 – is just fast enough to allow a safe climb out at V_2 with one engine failed and at the same time just slow enough to stop before the end of the runway. Length of runway, height of the obstacles to be overflown just after take-off, headwind, atmospheric pressure and temperature vary this maximum weight, furthermore a wet, slippery or snow-covered runway would reduce performance.
3   Tyre speed limit, peculiar to Concorde. The limit here is 250mph (220 knots) and can be a factor when the ground speed on take-off is high, due

to heavy weight, flying from a high-altitude airport or when there is a tailwind.
4   Noise abatement, also peculiar to Concorde. Special graphs were drawn to show the maximum weight for particular runways at New York from which Concorde would not exceed a given noise reading.
5   Centre of gravity position. Provided all the tanks were practically full, save tank 11 at the rear, it was allowable to have the centre of gravity at 54% rather than 53.5% (percentage of the root chord of the wing which, see Appendices, is 90ft or 27m long). This 5½ inch rearward positioning of the centre of gravity improved the performance allowing around an extra tonne to be carried – provided limits 1, 3 and 4 were not exceeded.

Weight restrictions due to performance considerations could reduce range and/or payload. Should the wind suddenly change, causing a take-off weight restriction, an alternative runway might have to be requested. At Charles de Gaulle, this could entail a 4.2km taxi from 26R to 08L. If required a weight reduction can be achieved by burning off the excess fuel before take-off;

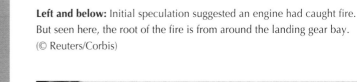

Left and below: Initial speculation suggested an engine had caught fire. But seen here, the root of the fire is from around the landing gear bay. (© Reuters/Corbis)

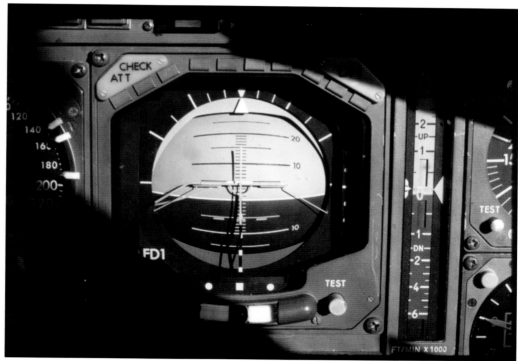

Left: The angle of attack indicator showing 3 degrees. On final approach it reads 14 degrees, so with a pitch attitude of 11 degrees, the descent angle is 3 degrees. At the final stages of the accident the reading was well over 16 degrees where there is huge aerodynamic drag. The pointer on the left side shows vertical 'g'.

Right: The artificial horizon on Concorde has a white 'bug', here set at $4^{1}/_{2}$ degrees and just visible behind the yellow flight director bars. On take-off it is set to θ_2 (typically 13–14 degrees) y a thumb wheel on the control column, on final approach to 11 degrees. The minimum pitch when supersonic is $-5^{1}/_{2}$ degrees, indicated by orange marks.

at idle thrust losing about 100kg per minute (compared with about 1,400kg per minute at full take-off thrust).

Using the actual take-off weight from the loadsheet, the 'V' speeds are confirmed. As already mentioned, V_1 is 'decision' speed. If there was an engine failure at, or less than, V_1, the take-off could (and had to) be safely abandoned. Once at V_1 or above, the take-off had to be continued, there being sufficient runway from which to get airborne. V_1 for Concorde fully laden was typically 160 knots.

The next speed is V_R (typically 195 knots) when the control column is pulled back to 'rotate', that is, pitch up the aircraft with respect to the horizontal. On Concorde with all engines operating, the rate of rotation required was 2 degrees per second to a pre-set angle of pitch θ_2 ('theta two' – the angle of pitch, or the the angle of the fuselage to the horizontal, that must be flown after

take-off, which is typically about 13 degrees). Approaching 250 knots the aircraft is pitched up again to maintain that speed, for the noise abatement procedure. Should an engine have failed at V_1, the rotation rate (still initiated at V_R) is slightly reduced. The aim now is to achieve V_2 (the engine failed climb speed, typically 220 knots) and a pitch of θ_2 simultaneously. V_1, V_R, V_2 and θ_2 are determined for each departure. Flight AF4590 from runway 26R, dry surface, calm wind, temperature 19°C (66°F), pressure 1008 millibars, was limited by the maximum permitted structural weight of 185,070kg; V_1 was 150 knots, V_R 199 knots, V_2 220 knots and θ_2 was 13 degrees.

Concorde generated lift at high angles of attack (in excess of 7 degrees) due to the formation of vortices over the wings. These gave lift at the expense of increasing induced drag (drag due to lift). The slender delta wing, at even higher angles of attack (over 16 degrees), did not stall conventionally but exhibited another phenomenon – it could generate more drag than there was thrust available. When thrust and drag were equal, the aircraft flew level, in other words it had a zero rate of climb. With maximum available thrust, the speed at which this occurs is known as V_{ZRC} – the zero rate of climb speed.

Following an engine failure at V_1 the subsequent climb profile is fixed using a compromise between climbing to clear immediate obstacles (using extra or 'contingency' thrust) and accelerating the aircraft to a speed well in excess of V_{ZRC}. Inability to accelerate was a major factor in the Gonesse accident. The following table gives the figures for V_{ZRC} at a weight of 185 tonnes:

V_{ZRC} (knots) 185 tonnes	3 engines	2 engines
Gear retracted	193	262
Gear extended	205	>300

Concorde crews were very conscious of this phenomenon.

For optimum climb performance on three engines, Concorde had to be flown with zero side-slip, in other words straight into the oncoming airflow. There was a side-slip indicator beneath the compass (horizontal situation indicator). If a left-hand engine failed, the aircraft would point to the left and, without rudder input, slip (or crab) to the right. Application of right rudder stopped the slip, but due to asymmetric thrust there was a small residual turn to the left. This was arrested by applying about 2½ degrees of right bank

The centre of gravity for take-off was usually at 53.5% along the wing root chord (see above). 1% is about 11 inches (roughly 27cm). The position of the 'zero fuel centre of gravity' was entered on the computer on the Flight Engineer's panel. This figure was derived from the disposition of the payload, pantry (meals) and crew. The computer then summed the fuel from the 13 tanks (numbered 1 to 11 – there was a 5A and a 7A) and produced the actual centre of gravity. Most often this was at variance with the 53.5% required. If it was forward of 53.5% then fuel was transferred to the rear tank 11; to the rear of 53.5% then fuel had to be transferred forward. In this condition all the forward tanks were full so there was space to complete the transfer only after the engines had used some fuel during taxiing. Therefore a pre-calculated amount of fuel had to be burnt off. Should the centre of gravity have been in excess of 54% on engine start-up, it was allowable to use 54% for take-off.

Top: To fly with minimum drag the side slip should be zero. The indicator moves to the right if a left engine loses thrust and vice versa. The sensing vane is beneath the flight deck

Bottom: The zero fuel weight and zero fuel centre of gravity are entered before the flight, on this occasion, 88.3 tonnes and 52.84%. Just above, displayed in lit digits, is 56.5% – the actual c of g computed from the quantity of fuel in each tank. The vertical instrument to the left shows the range of c of g available at a given Mach number, indicated by the yellow bugs 'fwd' and 'aft'. The instantaneous weight of the aircraft and the fuel remaining is displayed as well. The fuel was being transferred reward (see selector) when this picture was taken.

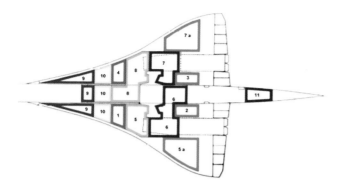

There are 13 fuel tanks; numbers 9, 10 and 11 are used to trim the aircraft, numbers 1, 2, 3 and 4 are the feeder tanks for their respective engines. Tank 5 was ruptured during take-off from Paris.

As has been alluded to already in this book, a rearward centre of gravity gave 'flap' effect at the expense of making the aircraft less stable. Stability with full tanks (apart from 11) was not compromised. The flap effect was put to use since it improved a performance limited take-off weight by about a tonne.

On the morning of 25 July 2000, the number 2 engine of F-BTSC had an unserviceable thrust reverser unit. Although Concorde was allowed to depart, once the unit has been safely locked, it incurred a performance penalty, which reduced the range. The Captain requested that the unit be changed in spite of the delay. It all became part of the intense media speculation over what caused the ensuing tragedy.

The following day, BA resumed its Concorde flights. Even tiny events of no consequence became the focus of enormous press interest. Because the cabin crew detected a faint smell of kerosene in the galley, one New York-bound Concorde diverted to Gander in Newfoundland. Before the accident, such a diversion might not have been warranted. The media was ready to pounce and the Concorde crews knew it.

Questions were already being asked. Had the recently reported hairline cracks in the wing caused the disaster? Did the last-minute change of the thrust reverser cause an engine to catch fire? Study of the photographs taken of F-BTSC on departure revealed a burning wing. How had a fuel tank been penetrated?

The subsequent investigation was to produce four reports and take 18 months to complete.

On 27 July the Bureau Enquêtes Accidents (BEA) issued the first of 14 bulletins or *Comuniqués de presse*. Paraphrased and translated from French it stated:

Shortly after V_1, the 'Tower' (Control Tower) reported seeing flames coming from the rear of the aircraft. Engine number 2 appears to have lost thrust (noted by the crew) followed later by number 1. The undercarriage would not retract. Speed and altitude remained constant and the flight lasted for about a minute. After banking sharply to the left, the aircraft crashed. The remains of tyres were found on the runway, debris was found along the flight path and in a small area at the crash site.

A Preliminary Report would follow at the end of August. The bulletin made two further points. The first mentioned that British, German and American investigators would be included under the auspices of the BEA. The British equivalent to the BEA is the Air Accident Investigation Branch (AAIB).

The second point acknowledged the role of the French judiciary, whose task was to take action should the law have been broken. In France any evidence is 'owned'

The piece of tank 5, sealed in a polythene bag which prevented the investigators getting sufficient access.

The metal strip which fell from the DC10 shortly before F-BTSC's departure. The shape of the strip prevented it from being flattened by the tyre. Instead it was 'locked' at right angles to the tread and cut the tyre from shoulder to shoulder. The wheel's direction of travel was as if it were coming from the top of this photo.

by the judiciary which can take advice from the investigators. In the UK the evidence is 'owned' by the investigators who work in parallel with the Coroner or Board of Inquiry. The investigator's remit is not to apportion praise or blame. By preventing early analysis of some parts of the wreck, the AAIB felt that the French judiciary had impeded the investigation. This represented a contravention of ICAO (International Civil Aviation Organisation) Annex 13 to which France is a signatory.

On 28 July the second bulletin appeared. Paraphrased it stated that:

> ... debris found on runway 26R came from the Concorde's left-hand side including remains of two[*] (of the four) tyres from the left gear leg. No engine debris was found and evidence suggested that the fire was external to the engines.

On 30 July the third BEA bulletin released more details, including that of the discovery of a portion of fuel tank on the runway. The judicial authority retained this piece. The polythene bag in which they had sealed it became opaque from being handled. Actions like this, said the AAIB, had delayed the investigation.

* Subsequently proved to be just one.

A tyre burst causing a fuel leak, which was somehow ignited, now appeared as the most likely explanation for the accident.

The fourth BEA bulletin, on 1 August, defined the seven areas of investigation for the Commission of Enquiry (presided over by M Alain Monnier):

- site and wreckage
- aircraft, systems and engines
- preparation and conduct of the flight, personnel information
- flight recorders
- aircraft performance
- witness testimony
- examination of previous events.

From the fifth bulletin on 4 August came the announcement that a metal strip about 40cm long, not belonging to the Concorde, had been found among the debris on the runway.

Was this the cause of the tragedy? How did such an object come to be on the runway in the first place?

The answer to the first question appeared to be 'more than likely'. The second question was not answered immediately, but started a debate about runway inspections. The requirement was three times

The cut in the tyre corresponded to the shape of the metal strip. Half the circumference of the tyre became detached and hit the tank above.

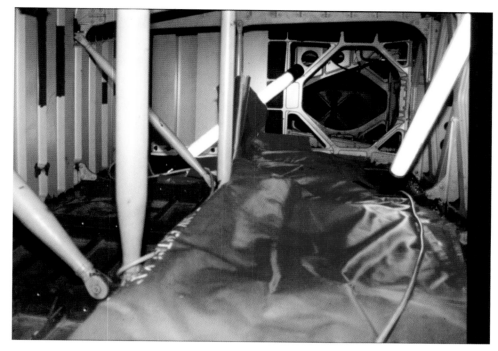

Left: The interior of tank 5; the black ribbing on the left shows the bottom of the tank. The red mat is for the maintenance engineer.

Right: The portion of tank 5 found on the runway after the accident showed no signs of having been impacted. It measured 30cm x 30cm which allowed an initial leak rate of 100 litres per second.

daily. Then it transpired that, due to a fire practice on 26R (runway used by F-BTSC) and 26L, the second inspection had been delayed.

On 16 August the eighth bulletin announced that Concorde's Certificate of Airworthiness had been suspended:

> … Le BEA et son homologue britannique, l'Air Accidents Investigation Branch, ont émis une recommandation de sécurité visant à la suspension des certificats de navigabilité des Concorde…

On the grounds that such an event might easily happen again:

> …le BEA a déterminé que c'est la destruction d'un pneu – événement simple dont on ne peut affirmer qu'il ne puisse se reproduire …

The previous day BA had been warned by the CAA of the imminent suspension of the Certificate of Airworthiness. For everyone involved with Concorde, having operated it safely for nearly 25 years, this was a bitter blow. A New York-bound BA Concorde flight had just commenced taxiing for take-off at Heathrow when it was recalled to the stand. John Tye, the First Officer, remembers acknowledging the instruction to return and the gloom that ensued at what seemed to be Concorde's finale.

There had been occasions involving tyre deflations, but following the incident described below, these had been satisfactorily addressed. By 1993, following further modifications, tyre incidents were practically eliminated.

What had been the most notable tyre burst happened to Air France Concorde F-BVFC on 14 June 1979 at Washington Dulles. The cause was not established. During taxiing before take-off, a main wheel tyre deflated probably due to faulty 'fusible plugs'. These were fitted to prevent an overheated tyre from exploding. The neighbouring tyre on the same axle now bore twice its normal load for the whole of the take-off run (there were two axles and four wheels on each of Concorde's

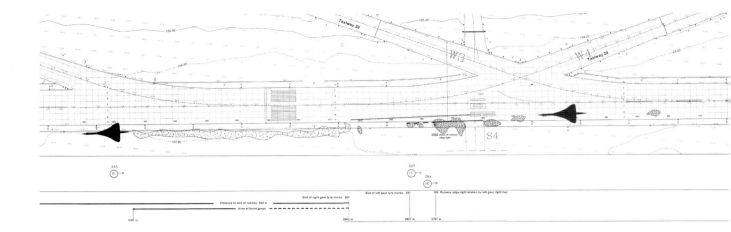

Plan showing runway 26R at Charles de Gaulle, Paris. Where the tyre bursts is shown by a zig-zag line, followed by shading indicating the soot deposit left by the flames. At 1,957m evidence of an explosion (which detached a piece of runway concrete) can be seen. This was the double engine surge which started the deviation to the left. Ultimately (but not shown on the plan) the aircraft hits a runway light before becoming airborne.

two main legs). According to TSS the overloaded tyre should have coped. In this case V_R (rotate) speed would have been about 190 knots. On take-off, the slender delta wing did not give lift until there was a distinct angle of attack. On take-off the conventional swept wing starts to give lift as soon as the speed builds. During the take-off run, the wings of a B747 start to give lift at a speed well below V_R (rotate) and curve upwards; this reduces the weight that is carried by the wheels. With Concorde at V_R, the wheels were momentarily forced into the ground so bore a force somewhat greater than the weight of the aircraft. This resulted in the extra-loaded tyre bursting. Debris from the wheel rim penetrated the tanks, damaged some hydraulic piping and caused a fuel leak of 6 litres per second (one tenth of the average rate of the Gonesse accident). No fire followed. (At full thrust and with reheat an Olympus engine uses over 7 litres per second.)

To prevent an incident similar to that at Washington, Concorde was fitted with stronger tyres and wheels, improved protection for the hydraulic pipes and a tyre deflation detector. The detector worked by sensing the twist in the undercarriage bogie beam. Should a tyre have deflated in the speed range of 10 to 135 knots, this failure would have been signalled to the crew and the take-off stopped. This system avoided exposing the neighbour of a deflated tyre to the major part of the take-off run.

When the Certificate of Airworthiness was suspended, Claude Freeman (BA Engineering Manager, Concorde) described his feelings in a television documentary: 'It was only a piece of paper … but it did represent Concorde's right to fly.' Was suspension inevitable? If a similar incident occurred to the world's most common airliner, the Boeing 737, instead of to the world's least, would the B737 have been grounded?

Concorde had been flying for almost 25 years; however, the 84,000 flight cycles achieved equalled those flown by all the Boeing 737s in a matter of weeks. Even if it were considered statistically impossible to repeat the crash circumstances, the authorities had little option than to call for the suspension of the C of A until appropriate modifications had been carried out.

Some argued that the lower wing skin was too thin to withstand impact and that the expense of strengthening it would be prohibitive. Now, they suggested, would be a good time to retire Concorde. Others pointed out that self-sealing tanks had been used in the Second World War. Nobody doubted that a remedy was possible, but would it be cost-effective?

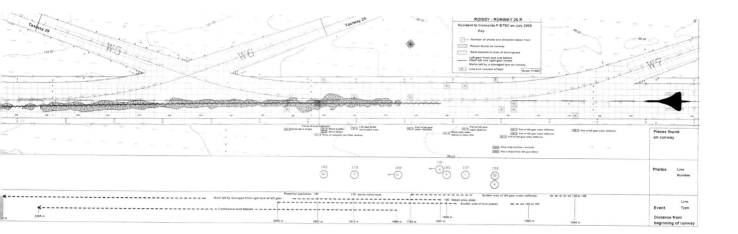

At first Air France did not appear as eager as BA to return Concorde to service. Soon after the grounding, Jean-Claude Gayssot, the French Transport Minister, said this was not the end of supersonic flight. Even before the crash, France had set up a commission to study an environmentally acceptable successor to Concorde. European or worldwide funding would be needed, he noted. Was this in preparation for announcing an end to Concorde with the connivance of Air France? The airline's attitude probably had more to do with its being in shock. Soon both sides of the Channel became equally dedicated to returning Concorde to service.

On 31 August the preliminary report was published. Its summary said:

> During take-off from runway 26 right at Roissy Charles de Gaulle Airport, shortly before rotation, the front right tyre of the left landing gear was damaged and pieces of the tyre were thrown against the aircraft structure. A major fire broke out under the left wing. Problems appeared shortly afterwards on engine number 2 and for a brief period on engine number 1. The aircraft was neither able to climb nor accelerate. The crew found that the landing gear would not retract. The aircraft maintained a speed of 200 knots and a radio altitude of 200 feet for about one minute. Engine number 1 then stopped. The aircraft crashed on to a hotel at La Patte d'Oie in Gonesse.

The preliminary report revealed that extra baggage had been loaded but not properly accounted for; the 'ground' copy of the loadsheet could not be found. Only 800kg of the two tonnes of fuel loaded for the taxi had been used. The extra baggage and fuel made the aircraft at least one tonne over structural weight for take-off, although this resulted in a negligible difference to performance.

The centre of gravity on engine start was 54.2%, which after taxiing had moved forward to 54%. The report told how data from the recorders had been retrieved. Ten seconds before the start of the take-off run, the cockpit voice recorder (CVR) reveals that the tower informed the crew of a tail wind (easterly at 8 knots) and cleared AF4590 for take-off. If this were a steady wind, then due to the tyre speed limit, the aircraft's performance weight was too great by about 5 tonnes. The crew did not audibly discuss this. In reality the *average* wind was very light from the north-east. The report said:

> At 14 h 44, the average wind at the threshold of runway 26 was 020°/3 kt and 300°/3kt at the threshold of runway 08. [*the reciprocal runway*]
>
> All goes normally to V_1 – 33 seconds after start of take-off. Six seconds later, at 175 knots, there is a noise and a second later a change in background noise – the ignition of the fire and engine surge (akin to a backfire). One second later, at 185 knots, the rotation is commenced.

The report shows pictures of the 43cm x 3cm strip of metal that cut the tyre from shoulder to shoulder, and the piece of damaged tyre itself, measuring 100cm x 30cm and weighing more than 4kg. The profile of the cut corresponds to that of the metal strip. On visual inspection the metal appeared to be a light alloy. There is no mention of where it could have come from. The photograph of the 30cm x 30cm portion of fuel tank found on the runway shows that it suffered no impact damage but it was slightly bowed outwards. The mechanics of its ejection from the lower wing surface are not discussed.

There is a runway diagram showing where the debris was found, the tyre marks of the left-hand undercarriage and trail of soot from the fire. Both the strip and the tyre debris were found together on the north side of the runway. This puzzled Alan Simmons, an investigator from the AAIB. Had someone put them together?

Why the rotation was commenced some 15 knots below the calculated V_R has not been satisfactorily explained, even in the final report which was published in January 2002.*

Immediately after rotation, there is evidence of an explosion when a piece of concrete (10 x 25cm and about 1cm thick) was detached from the runway probably where the two left engines surged simultaneously. The increasing angle of attack changes the airflow pattern under the wing. This caused the hot gases to be ingested into the engine via the auxiliary intake, behind the main intake in the floor of the ducting to the engine (see Appendices). The double surge gave the aircraft an impetus to the left, causing the left gear to strike a runway edge light before becoming airborne.

In conjunction with the physical evidence, the 'traces' of the flight are read and are shown in the report. Each 'trace' plots, with respect to time, a parameter: airspeed, pitch angle, engine thrust and other data. When the Black Box or Flight Data Recorder (FDR) reads 97602.5 seconds, which equates with 41 seconds from the start of take-off, there is a sudden decrease in acceleration. Fractionally later there is a strong lateral acceleration to the left (the double engine surge), which is countered with the application of right rudder. On rotation, the runway's centre line becomes progressively obscured. Without visual cue the Captain maintains runway heading on his compass while the aircraft continues its drift to the left from the impetus caused by the surges. The final report says the lateral acceleration sensed on the flight deck is less than that at the centre of gravity, which helps to explain the lack of track correction.

Within three seconds of the fire igniting and within one second of rotation, the control tower tells the crew there are flames behind them.

Two seconds later (45.5 seconds), at 195 knots, nose wheel off the ground and with less than 2,000m of runway remaining, from the CVR the Flight Engineer possibly says 'stop' [the take-off]. Perhaps he thought there had been a double engine failure, because when he announces 'engine failure' there is a hesitation about which engine has failed. Then he announces 'shut down number 2 engine'. (The standard procedure would be for the Captain to ask the Engineer to shut down an engine once at a safe height. On this occasion, the situation probably appeared to require instant action.)

One second later, the Captain asks for the Engine Fire Procedure. The fire warning for the number 2 engine sounds, supporting their diagnosis. In fact the heat of the fire outside the engine nacelle has set it off. Meanwhile, the number 1 engine recovers to give almost full thrust. The aircraft struggles up to 200ft above the ground. The speed barely reaches 210 knots, 5 knots above V_{ZRC} with three engines operating and gear down. If the number 2 engine had been operating they might have been able to accelerate despite fire damage but with no way of putting out the fire.

---

* At the end of this chapter there is a possible explanation. The rate of rotation was slow, about 1 degree per second, possibly in compensation.

In the Northern Hemisphere the wind veers and increases with height. On this take-off, although the wind direction was momentarily due east, the surface wind was generally north-easterly and very light. So on climbing to the west, their airspeed was not augmented with an increasing headwind, if anything the reverse.

The selections made on the flight deck can be interpreted through analysis of the sound signature of the recorded 'clicks'. For instance, the pulling of the fire handle can be heard 58 seconds after start of take-off. This occurs just after an unknown source, presumably another aircraft, has told them the flames are large and do not seem to be coming from the engine.

The landing gear does not retract. This and the low airspeed (le badin) are of great concern to the crew. 'Le badin, le badin,' calls the First Officer. A toilet smoke alarm sounds; smoke in the cabin? Air Traffic Control offers them 'an immediate return to the field'. This would involve a right turn for a landing to the east on the northernmost of the three runways at Charles de Gaulle (runway 09). The First Officer acknowledges. From 88 seconds after take-off the engine fire alarm sounds continuously and the terrain warning system urges them to 'pull up, pull up'. Then 100 seconds after the start of take-off, the number 1 engine, which had staged a recovery, surges and fails. The crew elects to try for Le Bourget, which by then is less than 2km ahead and slightly left.

At 205 knots with the gear down their speed is 100 knots below the two-engine V_{ZRC} for their weight. The aircraft can only decelerate. The rudder loses effectiveness, the aircraft banks to more than 90 degrees to the left and pitches up – the loss of fuel has moved the centre of gravity aft. It turns almost through 180 degrees. In an attempt to level the wings the crew probably throttled back the two right engines. With very little forward speed the aircraft impacts, breaks up and burns. The accident is not survivable.

A Japanese passenger took two photographs of the fire from a Boeing 747 waiting for F-BTSC to take off before crossing runway 26R. By coincidence Jacques Chirac, President of France was on board. Could the 747

The almost intact central and right-hand flight deck panel reveals some of the last second readings. The nose and visor selector (to the right of the top row of four dials) is in the landing position.

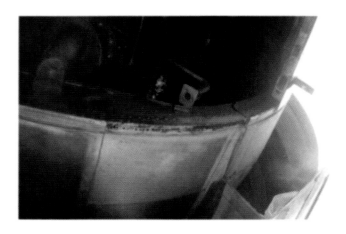

The missing lower left wear strip from the thrust reverser of engine 3 of DC10 registered 'N 13067'. An Air France B747 took off between F-BTSC and the DC10.

have triggered the early rotation? It was too far away to be a factor, on the last of a group of three taxiways.

Once on fire what else could the crew have done? The final report says that if the crew had tried to stop at the first indication of a problem at 183 knots or when the engineer may have said 'stop' at 196 knots, they would have overrun the runway at 75 knots or 115 knots respectively. Maximum braking on seven wheels and reverse thrust on three engines was used in the calculations. Neither course would have improved their chances of survival. The crew could do nothing more than they did; circumstances, through a set of cruel coincidences, had blocked all avenues of escape.

The report gives a résumé of the members of the crew, their qualifications and licences held. Christian Marty, at 54, had been a Concorde Captain for two years. Previously he had flown the Airbus A340 and before that a variety of mainly short-haul airliners. He had an adventurous streak. In 1982 he had crossed the Atlantic on a windsurfer. He had refused to sleep on his support boat, preferring to be strapped to his board; therefore he could truly say that he had spent the entire crossing on a windsurfer. On another occasion he flew over a volcano in a hang glider.

Jean Marcot, 50, had been a First Officer on Concorde since 1989. Rather than bid for a command

on another type of aircraft, he had elected to remain in the right-hand seat of Concorde. In theory his licence medical had expired eight days before. This oversight was more administrative than careless. The regulations had recently changed – previously a medical certificate had remained valid to the last day of whichever month it was due to expire. In July 2000 it was only valid six months from the date of the last medical. In November 2000 the rules reverted. He was an instructor on the Air France Concorde simulator.

Gilles Jardinaud, 58, the Flight Engineer, had had just over three years on Concorde.

The preliminary report confirmed the decision to keep Concorde grounded, with this final paragraph:

The Certificates of Airworthiness of Concorde be suspended until appropriate measures have been taken to ensure a satisfactory level of safety as far as the tyre destruction based risk is concerned.

On 4 September 2000, the 10th bulletin announced that the metal strip that had caused the tyre burst had fallen from the thrust reverser mechanism of a Continental Airlines DC10. This flight had left for Newark in the United States some minutes before the ill-fated Concorde. The author of the bulletin was at pains to emphasise that Continental had co-operated fully with the investigators.

Several major questions remained unanswered. How was the fuel leak caused? What was the source of the ignition? Why did the landing gear not retract? Did the missing 'spacer' in the left main gear cause Concorde to track to the left?

In each main undercarriage on Concorde there were two 'shear rings' whose purpose was to keep the wheels running straight. The rings were kept in position by a 'spacer'. If the spacer was missing and the undercarriage was retracted, the lower (outboard) shear ring was kept in place by gravity. However the upper (inboard) shear ring, unsupported by the spacer, would fall a little each

When the wheels are down, the gear doors are closed. After take-off, the gear doors are opened before retraction. Failure of the left door to open may have prevented the undercarriage retracting.

time the gear was retracted. Once the ring was displaced the bogie beam could become misaligned by up to 2 degrees in either direction. The bogie beam of the left undercarriage assembly carried wheels 1, 2, 5 and 6. During the accident investigation a BA engineer noticed that the spacer had been missing from the crashed Concorde. How significant was this?

Captain John Hutchinson along with other former Concorde flight crew, suspected that the lack of the undercarriage 'spacer' was another factor in causing the aircraft to track towards the left side of the runway. For most of the take-off run there were no discernible rubber deposits from the left gear and none from the right gear. There is, however, a photograph in the Final Report showing rubber marks from the three remaining wheels of the left undercarriage leading up to the broken runway lamp. The track of the two left tyres (numbers 1 and 5) on the left gear is clearly visible in the photograph.

To their right, in the direction of take-off, the marks from the flaying remains of number 2 tyre are visible ahead *and to the right* of the rear right-hand tyre (number 6) track. This suggests that the left bogie was twisted to the right therefore applying a force to the right. Being behind the centre of gravity, this force would attempt to push the tail to the right and the nose to the left. The left gear would drag more than the right, this too would increase the turning force to the left.

It seems the rubber deposits begin well after the start of the deviation to the left. The point where the rubber deposits appear could have been where the shear ring, normally kept in place by the spacer, was finally dislodged by the shaking of the flaying tyre. What caused the leftward drift?

Number 2 engine was shut down and the number 1 surged, ran down and took over ten seconds to restore itself to full thrust (see trace on p.159). In this time 1,000m were covered. With such a long period of asymmetric thrust, an application and holding of 20 degrees of right rudder would have been necessary. Initially 20 degrees of right rudder is recorded. For some

THE BLACK BOX RECORDING FOR AF 4590

Acceleration trace (below):

The blue line traces the acceleration along the runway.
At 97602 seconds the acceleration drops due to a loss of thrust
when the left engines surged. The green line shows lateral
acceleration which peaks at 0.22g to the left at 97605 seconds,
again caused by the loss of thrust from the left engines.

Rudder trace (opposite, top):

The yellow trace shows rudder position. At 97571, with the
plane having wandered to the right, the pilot applies left rudder.
The missing spacer from the left undercarriage therefore did not
pull the plane to the left. At 97603, to counter the loss of thrust

of the two left engines, right rudder is applied. Simultaneously
rotation is commenced (blue trace: despite annotation positive on
the chart = control column forward). The green trace shows roll
input. To keep the wings level, right bank is increasingly applied.

Thrust trace (Opposite, bottom):

The thrust can be assessed from 'p7' readings. At 97600
(39 seconds from start of take-off) the 'p7' of engine 2 drops,
indicating a loss of thrust when it surged. Two seconds later the
'p7' on engine 1 falters then recovers. Engine 2 starts to recover
but is shut down. By 97662 engine 1 surges again before
running down.

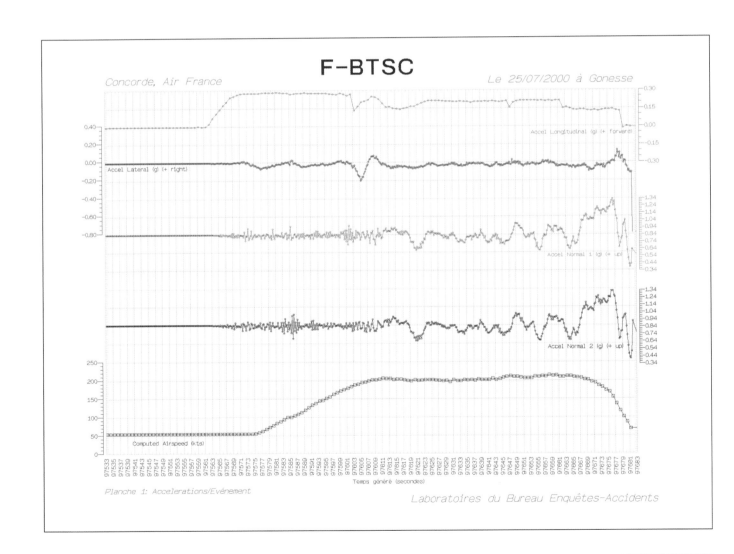

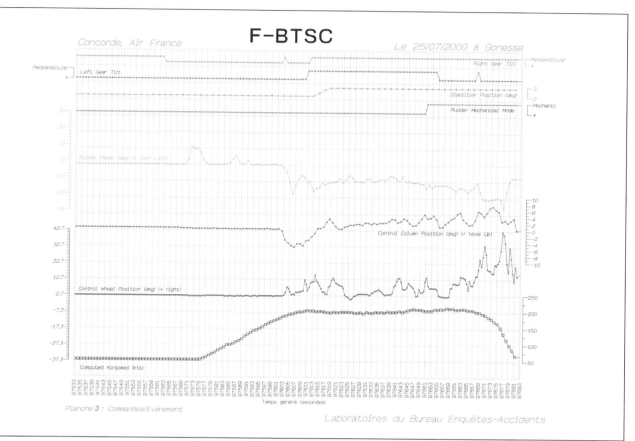

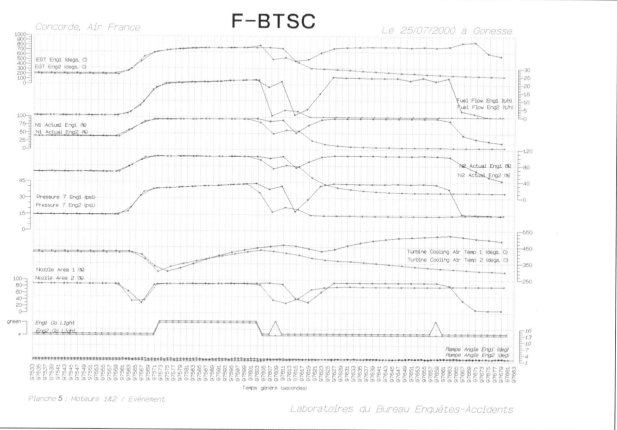

reason (autostab no longer detecting left yaw?) the rudder angle was reduced to, and held at, about 10 degrees. In the case of an engine failure when airborne it was recommended that sufficient rudder be applied to achieve zero (aerodynamic) side-slip. In this condition the aircraft would turn with wings level in the direction of the failed engine. To prevent this turn, about 2 degrees opposite bank is needed. If sufficient rudder only to achieve zero side-slip had been applied and the wings were level, as they would be with the main gear on the ground, a crab to the left would have resulted. The lack of the spacer may not have caused the initial deviation; but may have added to it just before lift off. By how much, if at all, is open to question. This subject was raised in a 'Discovery' TV documentary shown in summer 2003.

The wreckage of AF 4590 was transported to Dugny, Ile de France, for painstaking examination. (© Orban Thierry/Corbis Sygma)

According to Alan Simmons of the UK's Air Accident Investigation Branch, the tracks in the soot suggested that the left wheels had rolled over already deposited soot. Yet the wheels preceded the smoke. Had the wheels been soaked with fuel, which laid a damp trail on which the soot deposited differently? There was a fuel stain on the runway followed by a dry area then the deposits of soot. According to Ted Talbot (former Chief Engineer In Service Aircraft BAe) the piece of tank 5 hinged out, before breaking off. The hinge was inboard on the longitudinal axis. Fuel was first sprayed to the left – onto the runway, then downwards to the gear – not onto the runway, finally igniting – leaving soot on the runway. That would explain the observed evidence.

At some stage there might have been a leak from a damaged hydraulic pipe to the brakes causing a loss of 'green system', solely capable of raising the gear. 'Green' system was pressurised by pumps on the two left engines. There were two more hydraulic systems: 'blue' was

pressurised by the two right engines and was mainly dedicated to the flying controls. The standby 'yellow' system was pressurised by engines 2 and 4 and could be selected into parts of 'blue' or 'green' system. The French Judiciary had acquired the gear leg so the AAIB was not able to ascertain the exact state of the hydraulic pipes to the brakes. The AAIB's criticism of this practice appeared in the final accident report.

During gear retraction the main gear doors open first. Until it is up and locked, there is more drag than with it down. Could the First Officer have delayed gear retraction with this in mind? The second time the captain says, 'le train sur rentre' the co-pilot replies, 'le train ne rentre pas'. Is he saying the gear will not come up because of a malfunction, or because he is leaving it down? If the attempt is made, for some reason the gear does not respond. The red wheel light may have been on to indicate a brake overheat or that a flat tyre had been detected. This could have prevented selection of gear up.

In the documentary, some airport firemen suggest that they smelt tyre rubber and that the fire was already alight before 'SC' had reached the point where the infamous strip of metal was supposed to be. If this were true the tyre failure could be ascribed to the lack of a spacer allowing the wheels to 'crab', overheat and one to burst.

The evidence of the firemen is difficult to corroborate. Did they know where the piece of metal was with such accuracy that they could be certain that the fire had started before Concorde reached it? Suffice to say that the more verifiable evidence still supports the view expressed by the accident report. Tests were carried out on the about-to-be decommissioned F-BVFC in Toulouse to establish whether the lack of spacer was relevant. Apart from risking damage to what was now an irreplaceable museum exhibit, the test could not possibly reproduce the exact conditions of the day – the engine surges, the bursting tyre and so on.

Another conjecture concerns the question of the 'early rotate'. There was neither a call of 'rotate', nor of 'contingency'. Contingency (extra) power is automatically applied if one of the engines fails to give sufficient thrust. The flight engineer, if he spotted an engine problem once V_1 was passed, would manually select contingency thrust and announce that he had done so. It is possible that the captain, on hearing the words 'engine failure', reacted by rotating the aircraft. On rotation he simultaneously applied right rudder to keep straight and then, with the runway obscured by the nose, reduced the rudder deflection to fly with zero side slip, as discussed above. This might explain why the rudder is relaxed from 19 degrees and held at 10 degrees (see trace on p159). The crew found themselves in an entirely untrained-for set of circumstances, they heard calls referring to 'long flames behind you' amidst a cacophony of warning sounds which rendered themselves uninterpretable.

In such a scenario it is possible that a previously learnt 'conditioned reflex' took over. Captain Marty had started his career flying twin jets. On a powerful twin jet, V_1 (decision speed) is well in excess of V_R (rotate speed). It would be impossible to stop once airborne, so V_1 is limited by V_R. Every six months on the simulator pilots practice having an engine failure on take-off. The most difficult moment for this to occur is just as soon as the runway ahead disappears from view. So routinely every six months the candidate on a twin jet practises rotating and keeping straight following an engine failure. On four engine jets there is usually a gap between V_1 and V_R, typically of about 30 knots on Concorde. If a conditioned reflex had taken over, it might explain why when rudder was applied, the rotation was irresistibly initiated – 15 knots too soon.

In no way are these remarks intended to apportion blame, they are written to suggest why certain events occurred, and what can be done to prevent a similar accident. As already mentioned earlier in this book, the crew of 'SC' was faced by a cruel set of circumstances. Even with unlimited time it is difficult to suggest a better way of handling the incident; the crew of 'SC' only had seconds.

CONCORDE'S RETURN TO SERVICE

The pessimists declared that Concorde was a victim of its slender delta design. To meet all the aerodynamic criteria, the wheels, they said, had been placed too close to the engine intakes. Even if the tyres were stronger, they would only shed heavier pieces of rubber on bursting. To the optimist, historical precedence gave more hope. The will had been there to remedy the fault in Comet 1 for it to return as the Comet 4. A similar spirit had been shown in the case of the Space Shuttle returning to service following the Challenger disaster.

Meanwhile, BA was counting the cost of keeping its flagship inactive; a year's grounding would be sustainable, two probably not. The airline had always maintained that Concorde was profitable, and an indication of this came with the board's decision to take advantage of Concorde's downtime to start a £14 million cabin refurbishment. But without a Certificate of Airworthiness, Concorde was going nowhere, even if Sir Terence Conran was about to refurbish the cabin.

In early November 2000 a meeting was arranged at Gatwick to discuss the options for Concorde's future. Considering the meeting too crowded for useful discussion, Jim O'Sullivan (BA Technical and Quality Director) took five key people to a quieter room. Together Captain Mike Bannister (BA Chief Concorde Pilot), John Britton (Airbus UK, Chief Concorde Engineer), Alain Marty (Airbus France Chief Concorde Engineer), Hervé Page (Air France Concorde Engineering Manager) and Roger Holliday (Chief Airworthiness Engineer Airbus UK) came up with a plan of action which O'Sullivan outlined on a flip chart.

In essence it sought to:

- reduce fire risk by looking for leak and ignition sources
- study the behaviour of engines 1 and 2 and how hot gases cause engine surges
- validate performance
- check the hydraulics and the reason for non-retraction of the undercarriage.

In a lecture to the Royal Aeronautical Society (RAeS) at Farnborough in February 2002, John Britton spoke about the chain of events that led to the accident at Gonesse.

Starting with an incident that causes a tyre to break up, subsequent events can be traced through the chain depending on their seriousness. Thus tyre debris might cause 'structural expulsion' which would lead to either a major or minor fuel leak. If it is a major fuel leak and there is an ignition source, then this will lead to a 'catastrophe' after take-off. If there was either a minor fuel leak or no ignition source, then there would be no catastrophe. (Ideally, of course, there should be no tyre debris in the first place.)

According to the statisticians, the probability of a tyre burst causing a catastrophe was less than once in 10

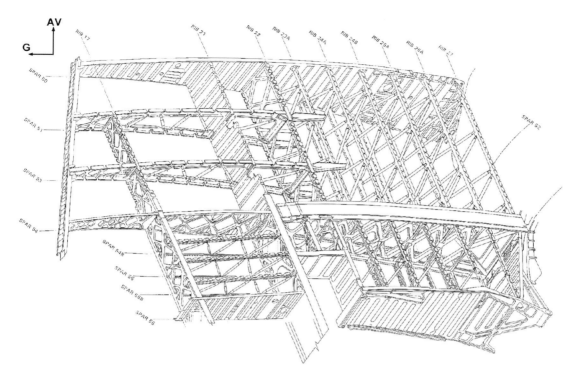

Plan of tank 5 in the left wing. Forward is to the top of the picture.

million flying hours. In 25 years the Concorde fleets accrued a total of 235,000 flying hours; at this rate a hull loss, resulting from a tyre burst, would occur once in over 1,000 years. The likelihood of a repeat of the Gonesse circumstances was minuscule. Was this the one event in 1,000 years that had merely happened early? That was a risk that the authorities would not take. There had already been too many incidents with tyres. Until the problem had been solved, Concorde would be grounded. BA and Air France set August 2001 as a target date.

The final report into the accident was released in January 2002. It describes some of the research that led to the remedy.

When the tank burst, the fuel was forced vertically downwards, initially at 100 litres per second before it stabilised at an average of 60 litres per second. A fuel stain on the runway before the soot deposits is evidence of this.

Showing no sign of having been impacted, the portion of tank 5 found on the runway must have been forced *out* of the tank. If so, it showed an entirely

new phenomenon. Had the 4½kg tyre segment caused an hydrodynamic shock to over-pressurise the tank? Possibly, but the fuel tanks of military aircraft penetrated by high-speed shrapnel had burst in a similar way.

Intense research followed. A computer model was assembled to simulate the mechanics of the incident. AU2GN, Concorde's aluminium alloy, is no longer produced but when some had been tracked down, a full-scale model of tank 5 was built. Centre d'Essais Aéronautique (CEAT) in Toulouse built a gun which could fire 4½kg lumps of tyre against the mock-up at 106m per second.

On 25 July 2000, a perfectly acceptable procedure had been followed to 'overfill' tank 5 to increase fuel quantity. As the aircraft accelerated, the fuel in tank 5 was forced to the rear, displacing any air towards the front. This effectively meant the tank was full in the vicinity of the external impact point, making it more susceptible to hydrodynamic shock. On the test rig, the impact caused the tank to be deformed inwards (direct mode), then outwards in an adjacent area in compensation (indirect mode). The rig did not burst but proved the principle.

Jim O'Sullivan (BA Concorde Engineering Manager) showing Captain Mike Bannister (Chief Concorde pilot BA) a piece of moulded Kevlar

The computer model, after being hit, shows a wave propagating along the lower tank surface. The skin then ruptures. Whatever the cause of the massive leak, a self-sealing device was required.

Kevlar is the name Dupont gave to the aramid rubber mix they had already developed. It is a light, black, flexible and very tough material suitable for bulletproof vests. Any tank that might be exposed to impact from an exploding tyre would be lined with Kevlar moulded to the interior contours. The Kevlar liner is suspended less than a centimetre above the tank floor, and each cell of the liner has a 4mm hole in it. As the outer skin of the aircraft is warmed during cruise at Mach 2, these holes allow convection of the warmed fuel away from the tank floor.

As the liners were being fitted, it was discovered that the tanks on each Concorde were slightly different in size. The liners had to be made to fit each individual aircraft – 'bespoke tailoring' as one engineer quipped. It was realised that there would be a weight penalty attached to the liners as well as to some 'unusable fuel'. This is the fuel that remains beyond the reach of the pumps, ie below the liner. The average rate of leak at Gonesse was 60 litres per second. The Kevlar liner would reduce such a leak to about half a litre per second. Only the most vulnerable tanks were to be lined.

The fuel leak patterns were studied when Concorde F-BVFB was flown to Istres near Marseille for ground testing. Accelerating on the runway to 175 knots, a leak (of an inert fluid) was simulated from an identical point to that in the Gonesse tragedy. Two other possible leak points were also assessed. Finding where the fuel went could reveal an ignition point and a flame stabilisation area – thought to be in the turbulent wake behind the undercarriage leg.

There were three likely sources of ignition – the reheat, an engine surge or an electric spark in the landing gear bay. The bay in question is just inboard of the engine nacelle and to the rear of the hole that had appeared in tank 5. From the distance between the fuel stain and the soot deposit on the runway, it was determined that ignition had occurred in less than a second after the tank was ruptured. All three ignition possibilities were tested at British Aerospace's Warton factory near Blackpool.

A full-size mock-up of the left under-wing section, the side of the engine nacelle, and extended gear leg were built and set up on a concrete platform. The airflow for the flame tunnel came from the exhaust of Warton's high-speed tunnel. The flow required was 106m per second (206 knots).

Fuel was released from the 'Gonesse hole'. Streaming over the open undercarriage bay to the rear of the wing, the fuel was lit by gas burners simulating the reheat. The fuel ignited, but the flames were unable to advance up the fuel stream traveling at 106m per second. The advance rate was less than 10m per second. This ruled out reheat as a likely source of the ignition. The French investigators believed that the fire might

Left: Every Kevlar section was specifically moulded for each individual Concorde.

Below: Mark Morley, BA engineer, fitting a piece of 'tailor-made' Kevlar in the cramped confines of a fuel tank.

have advanced in the slower moving air of the boundary layer (1mm thick) or through the various engine ducts. Neither theory proved likely.

To test whether the fuel leak was ignited by an engine surge, a 70-millisecond explosion was detonated and projected forward from the intake. A surge from the auxiliary intake of number 2 engine did ignite a fire that moved forward to the undercarriage bay. This cause was ruled out since the surge occurred after the fire had started; engines 1 and 2 then surged again, probably due to the aircraft's rotation changing the angle of attack. This change of airflow allowed hot gas to be ingested by the engine through the auxiliary intake.

A spark was generated in the undercarriage bay at the point where the brake cooling fan cable ran. On this occasion, a fire started in less than a second. It became stabilised in the eddies in the wake of the undercarriage

leg and side stay. The flame pattern closely resembled the photograph of Concorde on fire, much to the distress of some observers.

Before the crash the crew had tried several times to retract the undercarriage. The final report, published in January 2002, suggests that the door of the left wheel bay had failed to open. This has to happen for the undercarriage raising sequence to begin. The door could have been damaged by the fire or debris, but unfortunately after the impact the evidence was destroyed. The hydraulic power for operating the undercarriage is from the 'green' system. Engines 1 and 2 generate this system's pressure. No evidence exists to suggest that 'green' failed, until engine 1 ran down.

If the undercarriage could have been retracted the fire might have put itself out. But survival would have depended on how much damage had been done to the fuselage – which may have already been breached – and whether the number 1 engine could have been kept going.

A jet engine surge is a momentary reversal of flow through the engine, not dissimilar to a backfire on a car. The phenomenon is measured in milliseconds, and sometimes a flame can be seen in the engine intake. A surge can be caused by the engine demanding a greater airflow than is being supplied, and vice versa. Internal damage and foreign object ingestion are other causes of surge. The investigators wanted to study engine surges caused by the ingestion of neat fuel or of hot gases.

Accordingly, at Shoeburyness near the Thames estuary, an Olympus engine was mounted on a rig (which had lain idle for 25 years) fitted with a simple

Top: View of left undercarriage bay (gear down). The 'Gonesse' leak in tank 5 appeared just forward of the bay.

Bottom: A full-scale mock-up of the left undercarriage bay was constructed at Warton to investigate possible ignition sources. Here the cable to the brake cooling fans is being checked as such a source.

air intake. The tests proved that the engine performed as per the evidence from the Gonesse tragedy. Following a rupture in a lined tank, there is an emission of an initial 'slug' of fuel. This was represented by accelerating the flow to 100 litres per second, at a rate of 30 litres per second per second. The engine surged but recovered and coped well with the steady leak rate.

In January 2001, the first of two interim reports said that a 'spacer' had been found to be missing from the left gear of F-BTSC. An 'AO1' check had been carried out eight days before the accident, during which the left gear leg had been replaced. The spacer from the removed gear leg should have been fitted to the replacement unit. It keeps two bushes in place. Without their proper positioning, the whole gear 'truck' might wobble or shimmy, causing drag, thereby heating and damaging the tyres and impeding acceleration. Immediately questions were raised, particularly during a television documentary. Had the acceleration been sufficient? Had the aircraft been dragged to the left? More seriously, had the number 2 tyre already become overheated and damaged, so that on encountering the metal strip it burst even more violently than could have been predicted?

On 10 April the BEA addressed these points in its 12th bulletin:

- The brake temperatures were symmetrically warm during taxi.
- F-BTSC did not require right rudder to keep on the runway centre line before the tyre burst. In fact, a little left rudder was applied (due to a north-easterly breeze).
- The left gear did not leave rubber marks on the runway before the tyre burst, but tyre marks do appear on the concrete after it happened.
- The achieved acceleration of 0.268g was normal.

From this study the BEA concluded that the lack of spacer, though a bad oversight, did not contribute to the accident.

Early in 2001, the team working on Concorde's return to service were told that Michelin was putting the finishing touches to a tyre that promised to meet the most stringent requirements. The new tyre was almost impossible to burst, but should it do so the debris would be small enough not to be a threat to the fuel tanks above. It was called the Near Zero Growth (NZG) tyre, and was a development of the company's radial ply tyre.

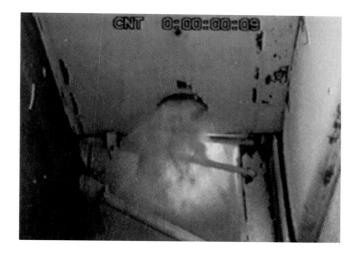

Top: This shows how the resulting flames stabilised themselves in the turbulent wake downstream of the gear leg and side stays.

Bottom: Not only is the power to the cooling fans isolated during take-off, but the cable has been protected with a woven steel mesh. Seen here along the top of the bay.

Concordes 'resting' at Heathrow. Without hydraulic pressure the two rudder surfaces are not aligned. Each surface is, on its own, capable of applying sufficient force in the event of an engine failure. (Mark Wagner/aviation-images.com)

Michelin had been making radial ply tyres for road vehicles since 1946. In 1981 the Mirage III became the first aircraft to be fitted with radial tyres. The NZG tyre is an evolution of the radial tyre, and the tyre for Concorde is reinforced with Kevlar. Its advantage is that it does not expand significantly with inflation and centrifugal force. The outer surface of the tread is not under tension. This makes it less vulnerable to foreign object damage – it is easier to cut an elastic band when it is being stretched than when it is not. Furthermore, the tyre is lighter than a reinforced cross-ply tyre. To prove these claims the new tyre was extensively tested.

Before trials on the actual aircraft, laboratory tests were carried out in the Michelin test centre and at the Centre d'Essais Aéronautique (CEAT) in Toulouse.

Tyres were subjected to extreme conditions laid down by the aircraft manufacturer.

There was an important point to establish – could the tyre run over a similar strip of metal and continue to perform normally? Two tests were devised, one a

The rear of the starboard wing without elevons, note the machined girder construction.

low-speed test and the other high-speed. The NZG tyre performed better than expected throughout.

Once, when the original cross-ply tyre was tested on a dynamometer, it burst and damaged the test equipment so badly that it took a week to rebuild. In contrast, the NZG tyre survived cut but inflated. In the end, only the Michelin NZG met the requirements of the damage resistance specification.

This was almost too good to be true. In case it was, BA kept up the momentum on fitting the tank liners and reinforcing the electrical power cable in the undercarriage bays. The tyre, the company said, could be seen as a bonus. Once the tyre had fulfilled its promise, it became an airworthiness requirement.

Inability to open the left main gear door (closed when the wheels are down) was considered the most likely reason for the non-retraction of the undercarriage. Non-alignment of the bogie beam could have been another cause. Note the position of the auxiliary intakes, the dark oblongs to the rear of the main intakes. They open to augment airflow at low airspeed and high thrust. (Mark Wagner/aviation-images.com)

In May 2001 the tyre was tested on an Air France Concorde at Istres. The regulations governing the anti-skid braking system had to be altered slightly to accommodate the characteristics of the new tyre. Luckily there was a heavy rainstorm while Concorde was at Istres, so on one of the three days of wet-surface performance tests the runway did not have to be artificially flooded.

Several modifications and stipulations were stipulated:

- Tanks 1, 4, 6 and 7 were partially lined, 5 and 8 fully.
- The cabling in the gear bay to the brake cooling fans was reinforced.
- There would be a change in crew procedures to cut the power to the brake fans during take-off and landing.
- Michelin NZG tyres would be fitted.
- The anti-skid braking protocol would be modified to give the aircraft identical performance to that with the old tyres.

The repainted 25-year-old test rig at Shoeburyness with an Olympus engine about to undergo surge tests.

- There would be a redesigned water deflector for the NZG tyre profile, without reinforcing steel wire.
- The tyre deflation detector would have to be serviceable for every take-off, ie. it would no longer be acceptable to fix it later.

On 17 July 2001 a BA Concorde under the command of Captain Mike Bannister, accompanied by Jock Reid, CAA Chief Test Pilot, prepared to take off at Heathrow for a test flight over the Atlantic. Members of the world's press were perched on a scaffold near the runway, reminiscent of Concorde's first commercial service over 25 years before. An ordinary subsonic aircraft took off. Then it was Concorde's turn, but there was a delay. Had some irritating gremlin lit a warning light? The reason soon became clear – the runway was being inspected for foreign objects. After the yellow inspection truck had left the runway, a distinctive roar could be heard coming from the threshold of 09R. Then that familiar shape was in the sky, reheats burning, undercarriage retracting, right turn commencing and the press applauding.

There were three main reasons for the flight. The first was to measure the unusable fuel remaining, held between the Kevlar and the lower skin. Second, it was to find out whether the liners had affected the temperature of the lower wing surface, normally kept evenly warm by fuel convection. Third, it was to monitor all the systems to see whether anything had suffered during the year's grounding. The plan was to fly on the normal Atlantic track, then at 20 degrees west turn north towards Iceland. Here, at about 55,000ft, they would find the relatively warm air (-50°C [-58°F] and warmer). The more northerly the latitude, at high altitude, the warmer is the atmosphere. At such temperatures the 'stagnation' temperature limits Concorde's maximum speed, no more than 127°C (261°F) on the nose. Using the readings from thermocouples fitted in the fuel tanks, the engineers would assess whether the circulation had been jeopardised.

Below: The tyre test truck was designed to check encounters with debris at low speed. Here the 'debris' is similar to that which caused the Gonesse tragedy.

Bottom: This was the result of the encounter; note the similarity to the 'Gonesse' cut.

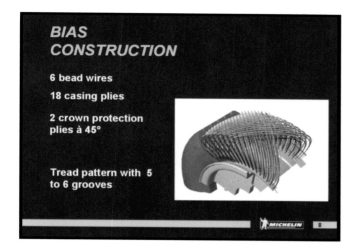

BIAS CONSTRUCTION

6 bead wires

18 casing plies

2 crown protection plies à 45°

Tread pattern with 5 to 6 grooves

The tyre with bias construction.

NZG RADIAL CONSTRUCTION

2 bead wires

4 casing plies

7 crown plies

1 crown protection ply

Tread pattern with 4 grooves

The form of the new radial tyre.

The new tyres and water deflectors were tested at Istres in a similar manner to this.

Concorde G-BOAF landed 3½ hours later at RAF Brize Norton in Oxfordshire. The world's press was there, and the low cloud, heavy rain and cold (for July) could not dampen the spirits. 'Concorde is back where she belongs, Mach 2 and 60,000ft,' announced a smiling Bannister. The flight had proved everything that it had set out to. On 6 September 2001 the Certificate of Airworthiness was reinstated.

Then came the 'dress rehearsals', with Concorde G-BOAF carrying 100 BA staff on a trip out to the mid-Atlantic and back, partly by way of reward for their work. The flight was an overwhelming success. In a matter of weeks, once three aircraft had been modified, services to New York would recommence. The date was 11 September 2001, and while the BA team were disembarking, news came through about the destruction of the World Trade Center and attack on the Pentagon. Would anybody want to fly now?

On 20 September, Brian Trubshaw's memorial service was held at St Clement Dane's Church in the Strand, London. The previous March he had died peacefully at his home in Gloucestershire. Jock Lowe, former Director Flight Crew BA and retired Concorde Captain, praised Trubshaw's life in test flying, and his contribution to Concorde. Trubshaw had been in regular contact with Mike Bannister, giving enormous encouragement to the Concorde relaunch effort. Sadly he died before Concorde had once again taken to the skies.

On 7 November, Lord Marshall, the Chairman of BA, his Chief Executive Rod Eddington, and BA guests, boarded G-BOAF for the first trip to New York since 15 August 2000. Valerie Grove of *The Times* was among the guests. She reported that Tony Benn was there to see them off. Benn told how he had fought for Concorde when the Treasury wanted to cancel it. He had also been on the test flight in 1970 equipped with a parachute. Then, with customary mischievousness, he told what he had said about the final 'e' of Concorde.

'"E" is for entente cordiale, "E" is for England.'

'But parts of it are made in Scotland,' had come the reply.

'Well "E" is for Ecosse,' said Benn, looking thoroughly pleased with himself.

Concorde's modifications may have cost £17 million, but the aircraft's return was due to the dedication and enthusiasm of the engineers, the pilots and the management of the two airlines. Would any other fleet of 12 aircraft with 25 years of service behind them have engendered such loyalty? With the exception of the Spitfire, it is difficult to think of another aircraft that has excited the imagination so much.

When Concorde first went into service there were two invitations: one to 'fly the flag, fly the future' and the other to 'fly Concorde and arrive in better shape'. The shape became an icon and Concorde through the three changes of BA livery retained elements of the Union flag. Other BA aircraft however, suffered their fins being decorated with 'ethnic' designs. On one occasion with full media coverage, former Prime Minister Margaret Thatcher famously draped her handkerchief over the ethnic fin of a model BA airliner.

Post-September 11, New Yorkers wanted to get back to normal as quickly as possible. On 7 November 2001, Mayor Rudolph Giuliani came to JFK airport to congratulate those whose hard work had restored Concorde to service. In so doing he praised the two airlines as Valerie Grove reported in *The Times*:

Sir David Frost, Concorde's most frequent traveller, sat at the back of the aircraft, having fallen into a light, refreshing doze. He woke in time to meet the outgoing Mayor of New York, Rudolph Giuliani, who boarded Concorde to welcome us all 'to New York City, the capital of the world,' adding:
'Do me a favour. Spend a lot of money.'

Top: Michelin should be justifiably proud of this superb tyre.

Middle: The Michelin NZG Kevlar reinforced tyre fitted to Concorde. Since the profile of the new tyre is different from the 'bias' tyre, a redesigned water deflector had to be fitted. This one is not reinforced with a steel cable.

Below: 17 July 2001, G-BOAF leaving LHR from 09R becoming the first modified Concorde to fly, the culmination of just under a year's dedicated work.

Captain Edgar Chillaud (left) with Captain Mike Bannister, Chief Pilots of Concorde Air France and British Airways respectively, at JFK, New York, on 7 November 2001 following the return of supersonic services.

He was addressing the right audience, as the flight was packed with captains of industry, chaps from Schroders and Allied Domecq and RM Rothschild. Dr Neville Bain, of Hogg Robinson, chairman of Consignia, professed himself thrilled to be back on Concorde, which gets him to New York for a late breakfast meeting and back home to London the same night. 'I'm no use at video conferencing, it's not the same,' he said. 'I need to eyeball people, and touch them on the elbow.' Also, you do not get Krug and caviar in a video conference.

As he spoke, the Air France and BA Concordes were parked nose to nose, a re-enactment of their pose at Washington on their first arrival in the United States in May 1976.

REFLECTIONS

In March 1984, Sir Frank Whittle was guest speaker at the Concorde Dining Society's annual event, held in the RAF Club, Piccadilly, London. He recalled his aspiration that a jet-propelled aircraft 'might carry mail across the Atlantic at 500mph.' In the 1930s this was regarded as crazily optimistic. As for being able to cross the ocean at 1,350mph in a 100 seat passenger aircraft he said, such a suggestion 'would have had me locked up'.

On 24 October 2003, after 27 years, the sun set on the first era of 1,350mph supersonic passenger flight. There is no successor in view and some contend that there never will be. I am not amongst them. The Concorde fleet still had (and indeed has) many 'reference flight cycles' available for use. These could have been translated into five or more years of service.

Indeed, following the tragedy at Gonesse, BA spent some £15 million on the modifications required to return all seven of its Concordes to flight and budgeted £14 million for the Sir Terence Conran Concorde cabin refurbishment, the 'interim' version of which was being fitted during Concorde's sabbatical. In the event only five were modified for a return to service and Air France paid proportionately less to modify four aircraft, choosing not to refurbish its cabins. Clearly, with well over £30 million being spent or budgeted for, at that stage no one thought that only two years of commercial service would remain. What went wrong?

On 10 April 2003, Air France announced that its Concorde fleet would, on 31 May, be withdrawn from service. BA would withdraw Concorde by 31 October 2003. According to Captain Mike Bannister, Chief Concorde Pilot BA, it was 'the right decision, taken at the right time'. The passenger-loads of both airlines were way below the pre-July 2000 figures. In particular Air France was suffering from an unofficial US boycott because of France's uncooperative stance over Iraq. On top of the knock-on effect that the conflict was having on air travel, the troubles faced by Enron, Worldcom and Arthur Andersen made it hard for businessmen to justify profligate spending on apparent luxuries like supersonic air travel. Perhaps to the joy of the company accountants, travel expenses diminished and, with interactive in-flight entertainment to alleviate the subsonic tedium, Concorde travel was no longer a corporate necessity.

Before the accident at Gonesse it was said that the new BA Engineering Director, Alan McDonald, was not as enthusiastic as his predecessors had been in continuing with Concorde. He had been concerned about the disproportionate man-hour costs, five or six times as much for a Concorde flying hour compared to that, for instance, of a B747 – in spite of Concorde flying twice the miles per hour of a subsonic aircraft. Furthermore new aircraft engineers were reluctant to become qualified on Concorde with its 'analogue' avionics.

'Au revoir Concorde', Paris CDG. The French reaction to the demise of Concorde appeared less emotional than that in the UK. Perhaps this was because the Airbus consortium, perceived as being mainly French, was continuing to build airliners. Nevertheless on 27 June 2003, André Turcat spoke at Toulouse during the handover to Airbus when 1,000 invited guests witnessed 'FC' stopping outside the hanger from which 001 had been rolled out 36 years previously.

In June 2000, a month before the Paris accident, it became apparent that the BA Concorde Major Checks would overlap one another. A 'Major' consists of a thorough service and testing of all systems and takes several months. Either the flying programme would have to be reduced or extra resources allocated to engineering. BA decided to stop charter flights.

When Concorde returned to service in November 2001, only the single daily New York flights and the once weekly Barbados flight were resurrected. At first the London to New York service departed at 10.30am, whilst the return schedule varied according to the number of serviceable Concordes available. With an aircraft available to 'night stop' in New York, the earlier and preferred service to London could be achieved. It was difficult to understand why the maintenance resource was inadequate in this respect; after all there were five modified Concordes, a number with which much more could have been done. For the last few months of operation the 10.30am departure to New York from London was changed to 6.30pm – 'to suit the US market'. Soon it became obvious that aircraft 'AA'

and 'AB' would not be modified with the Kevlar tank liners and would never fly again.

Captain Jock Lowe, former Chief Pilot of BA and Commercial Director of Concorde was very critical of the cessation of Concorde charters. In the early 1980s Lowe had very ably assisted Captain Brian Walpole, General Manager Concorde, during an earlier funding crisis (as described in a previous chapter). For his services in making Concorde profitable, Brian Walpole had been awarded the OBE. He retired in 1987.

It was not until 1993 that Concorde had a pilot as Commercial Manager again. Captain David Rowland* set up the Life Extension Programme, withdrew the Washington service and expanded the seasonal Barbados service. Stopping the Washington service, he said, was a difficult decision. Thanks to his efforts Concorde was predicted to stay in service until perhaps 2015.

From 1996 to 1999 Jock Lowe was in charge of Concorde's commercial operation. Thereafter, it was handled by 'marketing', like any other BA fleet. Even so, Lowe still monitored Concorde's progress. He was not, however, in a position to influence the decision to ground the aircraft after the Gonesse tragedy, a decision he questioned.

Sadly, the Paris accident proved that Concorde was not immune from serious mishap. Following one or two technical problems, or as declared by the media 'safety scares', *The Sunday Times* reported, on 15 December 2002, that 'The Jet Set gets nervous of flying Concorde.' BA explained that the low (50%) load factors were due to the economic downturn but the cost of the Concorde tickets appeared prohibitive to many, especially with the media reporting Concorde's every hiccup so gleefully. 'How can the Concorde fare be justified when pieces of rudder keep tearing off?' the media demanded.

---

* Captain Rowland presented Intelligent TV Video's (ITVV) comprehensive five-hour Concorde video (www.itvv.com)

'Concorde has two rudders, each capable of coping on its own,' was the reply.

'Is Concorde getting old?' the media persisted.

'Concorde is old in years but not in hours of use. Technologically Concorde is still way ahead of anything else,' came the defence. Yet, despite the number of people coming out in support of Concorde, they could not excuse a series of events that at best looked untidy and at worst heralded future failures; such as rudder delaminations.

The first rudder delamination happened to a BA Concorde during a round the world charter flight in 1989. The problem occurred when the rudder's aluminium alloy skin surface separated from the

honeycomb internal metallic structure. After this incident it became apparent that new rudders were needed, and new rudder jigs were constructed from which replacement rudders were manufactured. The problem appeared to have been solved. Then, in November 2002, G-BOAE lost a portion of a new lower rudder. The remedy needed remedying. In February 2003, at the same age in hours of flight as BA's first incident, an Air France Concorde experienced a similar problem and in early May 2003, Air France was to lose part of one more rudder, but by that time Concorde's fate had been sealed.

There were other incidents dismissible as 'teething' problems when the aircraft were new; but more difficult to justify after nearly 30 years' service. In April 2002, the British Chancellor of the Exchequer Gordon Brown flew on Concorde to the US for the G7 summit. En route G-BOAD suffered an engine surge. In consequence the aircraft was delayed by 30 minutes; an event considered newsworthy because of the high profile of both the passengers and the aircraft and, consequently, another nail in Concorde's coffin.

Engine surges during supersonic cruise are rare and even rarer at lower speeds. The author only experienced one such incident in ten years of piloting Concorde. A surge consisted of a sudden reversal of airflow through the engine. Usually there were a series of 'reversals', giving rise to the impression of flying through severe turbulence. The process started when there was a mismatch between air supplied via the intake and that demanded by the engine. A surge in one engine could induce a surge in its neighbour, making diagnosis of which engine was to blame difficult. At Mach 1.3 or faster, the drill, which had to be committed to memory by the crew, was to close all four throttles and pump the fuel forward, thus bringing the centre of gravity forward; with idle thrust, deceleration was quite rapid. The reduction in Mach number would result in a forward movement of the centre of lift that needed matching by a similar movement of the centre of gravity. This ensures that the elevons retained maximum authority. Although the attitude (angle of fuselage with horizontal) was

A doctored publicity photo depicting how a formation of four BA Concordes would have appeared in their final livery. In reality Concorde never flew such a formation in these colours.

A cirrus cloud formed by decaying vapour trails swirled into a familiar shape over the author's garden. Then, like the metal original, vanished from the skies.

kept constant, the deceleration gave the impression of starting a significant descent. Thereafter the drill turned into a troubleshooting exercise. The throttle computers were switched to their alternatives, the hydraulic systems responsible for moving the intake ramp doors were switched to the standby 'yellow' system and the 'lanes' (paths that the air intake computer used to vary the intake area) were changed over. Depending on why the surging started, there were appropriate courses of action, including a return to Mach 2 cruise. In the author's case, the cause was a broken 'stator' blade in the high compressor section of the engine, which had to be shut down. At Mach 0.95

(the maximum speed allowed on three engines) the fuel usage per mile was increased by about 25%. In this incident there was not sufficient range to reach New York, so we diverted to Halifax, Nova Scotia.

As already suggested, any such incident occurring post-Gonesse attracted disproportionate press scrutiny. Airbus (formerly British Aerospace and Aerospatiale) had worked hard helping Air France and BA return Concorde to service and once Airbus' high-profile offspring had been restored to health, Airbus could again proudly admit to its parenthood. To endure another major incident, or even a series of minor ones, threatened Airbus' reputation. Furthermore, Airbus

Concorde silhouetted on final approach to Heathrow. (Mark Wagner/aviation-images.com)

A view from 50,000ft of the weather heading towards Britain from the west. Here on the threshold of space the sky above is darker – although it is not quite dark enough to see any stars. Furthermore the atmospheric pressure is a tenth of its value at sea level. However with a differential pressure of 10.7 pounds per square inch the cabin 'altitude' is a little over 6,000ft. Whereas in the Boeing 747 battling against a turbulent 100 knot headwind beneath at 31,000ft, the cabin 'altitude' would be in excess of 7,000ft. In May 1980 Mount St Helens in Washington State on the west coast of America violently erupted, causing a layer of volcanic debris to be visible around the horizon which lasted for two or three years.

Concorde G-BOAF in the final BA livery. Concorde was not designed to be an icon, but to fly passengers at twice the speed of sound mainly across the Atlantic. To do this the wing was sculpted with subtle camber and twist which enhanced the ogee delta planform. There is a 'Mona Lisa' quality about Concorde – Leonardo's masterpiece poses the question of whether his subject is smiling or not, and that requires long scrutiny. With Concorde the question is why that shape does work so well.

wanted to concentrate on the launch of the 550-seat A380. Nothing should undermine its new position as the world's number one civil aircraft constructor.

Meanwhile Boeing, having opted out of the Sonic Cruiser project, was extolling the virtue of smaller, cleaner aircraft flying point-to-point.* This would be a better use of existing resources and avoid forcing everyone through a hub like Heathrow. Adam Brown, Vice-President of Airbus noted, in a radio interview, that the A380 would be the most fuel efficient airliner of all with the lowest levels of harmful emissions. According to BAA there could be an increase in Heathrow's passenger volume of up to 10 million per year (about 20%) without any additional investment in infrastructure through the use of such super-jumbos.

Concorde is all about speed and status. Since 1988, according to Jock Lowe, there had been a gradual decline in the numbers of clientele. To address this there had been a suggestion to rearrange the interior, having an extra luxurious front cabin while keeping the standard layout in the rear; a 'super-first' class and a 'super-club' class. With Concorde still profitable there was no serious pressure for change, so the modification was not made.

The author on the flight deck.

---

* The Sonic Cruiser was never built, being more of a design study about cruising at transonic speed. The cynics said it was designed to take the spotlight off the new Airbus A380.

This shows how part of the top rudder delaminated and broke off in flight. Alarming though this looks, the remaining rudder was designed to cover for such an eventuality.

A section of rudder showing the metal honeycomb structure which was revealed when the aluminium alloy skin delaminated. The coin with a diameter of 27mm, gives an indication of size.

Post-November 2001 this kind of modification might have helped. Had the Concorde operation been re-established as a 'division' or 'profit centre', then such changes would have been easier to implement.

Concorde served two groups, the 'premium' customers – top business people, sports and film stars – and those interested in experiencing Concorde. Each group encouraged the other. Following the return to service, charter flights could have been used as a way of rehabilitating Concorde and re-establishing her reputation. When Concorde's withdrawal was announced it became apparent that her reputation was largely still intact, at least among the general public – take-up of tickets soared. Clearly the market was there. Unfortunately it took a 'must-end-by-date' to realise it and by then it was too late, especially for Air France.

Since November 2001, if not well before, the Air France Concorde operation had not been profitable. In the past it had basked in *La Gloire du vol supersonique*. In early 2003, with privatisation in view, loss-making divisions of the airline would have to go. Then on 19 February 2003 Air France suffered another worrying incident.

A Concorde en route to Paris from New York experienced a failure, which demanded the shut down of an engine. The crew applied the appropriate drill requiring the closure of the 'HP' (High Pressure) fuel cock; but not of the 'LP' (Low Pressure) cock allowing fuel to circulate for system cooling purposes. There was, however, a fractured pipe downstream of the 'LP' but upstream of the 'HP' cock. Mercifully, the crew spotted the fuel loss in time and stemmed it by the closure of the 'LP' cock. There was no risk of a fire. They turned back to Halifax where they landed with minimal fuel. To prevent a repeat occurrence the drill was changed; the 'LP' cock would always be closed in similar circumstances.

With Concorde attracting bad publicity and losing money at the same time, Air France could see no point in keeping the aircraft flying. It would cease operating Concorde and stop paying the 'support costs' (payable to Airbus who held the Design Authority). As a result, BA would have to bear the full brunt of the support costs and if they were unable to, would be forced to end their Concorde operation. In 1962 the treaty between France and Britain to build Concorde forbade either country from withdrawing from the project unilaterally without paying its half of the total. In 1984 each country was deemed to be in balance with the other in respect of payments. In future there would be no

Overleaf: Concorde climbing at sunset, nose drooped. (© PCL/Alamy)

obligatory payments between the two countries if one of them wished to stop.

Airbus, with its reputation on the line and a new aircraft imminent, was not prepared to maintain support of Concorde at the old price. BA was not prepared to pay the new increased cost alone. It seemed to many that Airbus was losing faith in Concorde yet Captain Les Brodie, BA Concorde Training Manager, spoke of the confidence the crews had in the airframe, systems and back-up systems. There was plenty of spare capacity built in. Having flown Concorde on test, where emergency systems are exercised, Brodie's declaration of faith is a telling one.

BA wanted a year or more in order to bid farewell to its flagship. Air France chose 31 May 2003, some seven weeks ahead. Airbus agreed to continue to support BA for only six months. The final day was to be 24 October 2003. Concorde would visit former destinations on specially organised flights and a Grand Finale would be arranged at Heathrow for the last day of operation.

Thereafter, the Concordes that could fly would be flown to various museums, and those that could not, dismantled for haulage by surface. Once civil supersonic flight had become a finite commodity, all flights became overbooked. Was there no way that Concorde could be saved?

Sir Richard Branson, head of Virgin Atlantic, had always been an admirer of Concorde. He had in the past approached Air France with a view of obtaining at least three aircraft – about the minimum required to support a transatlantic service. Now he saw his opportunity. He noted that some of the BA fleet had been acquired for the nominal sum of £1 sterling. Surely he as a British taxpayer had some say in their disposal. (He conveniently

Looking west over a delta wing. When Concorde was operating it was possible to see the sun rise from the west. Sir Paul McCartney visited the flight deck on an evening flight from London to New York just as the sun was rising ahead of Concorde. It was suggested to him that he might like to call his next album 'Flying into Yesterday'. So far he has not taken up the idea.

A British Airways Concorde is towed to a hangar at Heathrow for inspection after it developed an engine fault on 15 July 2002. The plane, which was flying at supersonic speed over the Atlantic, was turned around after the fault forced the pilot to shut the engine down. (© Reuters/Corbis)

overlooked the £16 million paid by BA in the 1984 settlement and subsequent investment in modifications and refurbishment.) Concorde was a great British success story – if BA was unable to fly it, then Virgin would, and at reduced operating cost too. He would approach Patricia Hewitt, Labour MP and Trade and Industry Secretary, to gain financial backing. The government, remembering its predecessors' involvement, declined. Acknowledging that BA would have felt unhappy to see Concorde decked out in Virgin colours, Sir Richard suggested that the BA logo could remain on his newly acquired aircraft; on one side, with Virgin's on the other.

On 30 April 2003, Sir Richard appeared as a guest on the Richard and Judy television show on Channel 4.

On the telephone was Lord Marshall, Chairman of BA, who stated that Airbus would no longer support Concorde whoever the operator was. Sir Richard claimed that BA had instructed Airbus to say this, which Lord Marshall denied. When the results of a poll suggesting that 97% of the British population supported Sir Richard's bid were cited, Lord Marshall pointed out that if BA were to offer to keep Concorde flying it too would receive overwhelming support.

Sir Richard was not to be stopped so lightly. Uden Associates arranged for Sir Richard to narrate a Channel5 documentary also entitled *The Concorde Story*. Sir Richard was full of praise for Concorde; but also made a smiling reference to his own airline. On another occasion he was seen with an inflatable model of Concorde tucked under his arm, stating that as BA could not keep it up, he should be allowed a try. In the event his bluff was never called. Were there any other means of keeping Concorde flying? There were some who thought so.

Dermot Murnaghan interviewing enthusiastic Concorde passengers for the BBC. Seats were at a premium as soon as the announcement to cease Concorde services had been made. (Mark Gains)

A group of BA Concorde crew attempted a rescue. They extolled Concorde's virtues: its uniqueness, its power to inspire and, since the termination announcement, its earning capability. This 'Concorde Experience' market had the potential of lasting well over 18 months; it was a shame, they said, not to take advantage of it. The scheduled flights now resembled the former charter flights. As no flight deck visits were permitted whilst airborne (due to post '9/11' security measures), the crew offered to host visits to the cockpit after arrival, such was the dedication that Concorde inspired.

The group approached Rod Eddington, Chief Executive of BA. One of its proposals was to set up a Heritage Flight retaining an aircraft for special occasions – such as the Queen's Golden Jubilee fly-past. It would,

with no profit, have fended off the criticism BA was suffering for withdrawing Concorde. The termination costs however, some £80 million, would still have been payable. Airbus would not support this idea and it was dropped.

It was thought that Airbus might be more malleable if both BA and Air France were involved, so the second option was to form an alliance with Air France and perhaps even include Rolls-Royce. Although some Air France Concorde pilots liked the idea, Air France Chairman Jean-Cyril Spinetta was not interested, so this option failed as well.

The next hope was to operate Concorde in a wholly owned BA subsidiary company. This idea initially gained Rod Eddington's support but Airbus appeared intransigently set against supporting Concorde and it is doubtful whether the BA Board would have approved it. For the final option, that of a management buy-out, there was no chance of success. Even had Airbus approved, BA would not have done so, as it would have received less than from the other plans.

The company responsible for performing the product support function must hold the Design Authority (DA) of that product. In this case Airbus holds the DA of Concorde. If Airbus wished to withdraw support could the DA be held by another organisation? In essence the answer is yes; but Airbus would still have felt liable should there have been an accident attributable to a design error. Furthermore, Airbus would not have wanted to lose control of its intellectual property. The air intake control system, for example, was still regarded as being commercially sensitive. To whom did the intellectual property belong anyway? It could be argued that it was the taxpayers', or at any rate belonged to the governments of France and the UK. There were plenty of companies capable of holding the DA – Marshall Aerospace at Cambridge for one and BAe Systems for another (although BAe Systems would have found its loyalty divided since it itself is part of Airbus). It might even be possible for BA to hold the DA; after all, in early 2004 BA had as much experience of Concorde as any other organisation. The question was never resolved.

Concorde, whose slender delta and subtly cambered wings drew admiration from every angle, was about to be grounded for good. It must have been agonising for BA when arguably its greatest symbol was being forced out of the skies. The one aircraft that distinguished BA from all of the rest, apart from Air France, was to be removed. The 'exeunt' would need careful handling. To end supersonic services was one thing, but to botch up the overture to, and the Grand Finale itself would be truly unforgivable and damaging to BA.

A Concorde waits to take off as a Virgin Atlantic jet comes in to land at London Heathrow, during Concorde's final day of commercial service, 24 October 2003. Virgin Atlantic owner Richard Branson had tried to buy the Concorde fleet but been rebuffed by BA. (© Adrian Dennis/AFP/Getty Images)

GRAND FINALE

For Concorde's last day of commercial operation, 24 October 2003, BA had planned a spectacular celebration. Many of the world's press and television media congregated at Heathrow for this final Concorde photo opportunity. A grandstand was set up overlooking the northerly runway 09L/27R and a 'media stand' was constructed with a view of the final parking area. Three Concordes, landing in succession, were due to arrive at 4.00pm from Edinburgh, Heathrow (via the Bay of Biscay) and New York. All five modified Concordes were serviceable so in the event of an engineering problem there was plenty of cover. On the first day of commercial operations, 21 January 1976, two Concordes had behaved immaculately departing from Paris and London simultaneously. Would the same be true on the last?

In the morning of 24 October 2003 Captain Andy Baillie had flown G-BOAE to Edinburgh with a supersonic excursion out over the North Sea. Captain Les Brodie took over for the return to London. Both flights were full of BA staff and excited competition winners. In the afternoon the second Concorde, G-BOAF, captained by Paul Douglas departed London for a supersonic flight around the Bay of Biscay. Amongst the passengers were Concorde alumni – Captains Brian Walpole, Jock Lowe and David Rowland. Also aboard was John Cochrane, the co-pilot on Concorde prototype 002's first flight in 1969. At midday UK time (dawn Eastern Standard Time) the final transatlantic Concorde service departed New York with Captain Mike Bannister at the helm. This flight was scheduled to be the last Concorde to land at Heathrow. For those watching and waiting the weather was fine but, due to a brisk north-westerly wind, cold.

BBC presenter John Sopel and I were installed in the warmth of a temporary studio rigged on the media stand overlooking parked Concorde G-BOAC, which had visited Cardiff the previous day. Also waiting was former Concorde Captain John Hutchinson, scheduled to talk about the triple arrival in the afternoon. I asked him how he was and how he felt about the demise of Concorde. 'I am very well,' he replied, 'but I am sad, very sad. Personally I don't feel that BA should be giving up the aeroplane at all. It has got plenty of life left in it.' It was time for BA002 to depart from New York and my moment to commentate.

A monitor showed that Concorde BA002 had left BA's terminal at New York. The fire brigade arranged their traditional farewell – a water arch under which G-BOAG taxied. Would the water-ingestion cause the engines to flame out? Of course not and even if one engine were to, the automatic ignition system would instantly relight it. I previewed the take-off sequence hoping I had guessed the correct runway – 31L, I had. Soon Concorde was rolling along the same runway from which she had first taken off in October 1977. Prior to that first visit the clamour of protest had frequently exceeded any number of decibels produced by Concorde.

The noise abatement procedure for 31L had been carefully choreographed and test flown in Casablanca:

At V_R (typically 198 knots): Rotate to θ_2 (13 degrees) to be achieved over five seconds.

Once 500ft per minute climb achieved: left turn, applying 25 degrees bank over six seconds (the turn starts at about 50ft).

At a pre-calculated (weight and temperature dependent) time, typically 60 seconds: cancel reheat and retard throttles to a pre-set angle on the throttle quadrant; speed 240–250 knots (if speed is too low, < 235 knots delay the power reduction).

Once through a heading of 235°M (Magnetic): set full thrust rapidly (no reheat) and reduce bank to 7½ degrees, speed 250 knots.

2,500ft: throttles back to their preset quadrant angle, maintain 250 knots, increase bank angle to optimise tracking.

A crowd watches as the last of the BA Concordes lands at Heathrow. On the fence in the background a sign in French reads, 'Concorde, we love you'. (© Eva-Lotta Jansson/Corbis)

Over the ocean (5 DME Canarsie): slowly reapply full climb power over ten seconds.

If that is not enough to silence the reader, its application in Concorde silenced the protestors. The residents of Howard Beach, at the end of 31L, could have no case against a noisy aircraft if it did not fly over them.

For the last time Concorde G-BOAG performed the same manoeuvre. As she climbed, the reheats were switched off on cue. A helicopter (one of six), hovering to the north-east of Jamaica Bay, captured the whole sequence as Concorde continued her turn, climbing up and away towards the Atlantic and Great Britain. Air Traffic Control bade her farewell for the last time.

On 26 November 2003, Concorde G-BOAF is given the traditional farewell by the Heathrow Fire Brigade before flying to Filton and into retirement. (David Apps)

Some three-and-a-quarter hours later Concorde had crossed the Atlantic and was flying towards Ockham in Surrey at Flight Level 80 (about 8,000ft). Viewers of BBC2 heard the conversation between Senior Engineer Roger Woodcock, aboard BA002, updating Adrian Dolan of Air Traffic Control (ATC) in the Heathrow Control Tower with G-BOAG's latest position. Roger announced the revised route 'AG' would take; north over Heathrow to Hillingdon, east to the Thames Barrier, descending to 2,500ft, right-hand base leg to intercept the extended centre line of runway 27R, Greenwich, Canary Wharf, over the City with BA's London Eye on the left, over The Mall and Buckingham Palace and on to Heathrow; ETA 4.08pm local time.

Adrian Dolan said that it had been a pleasure looking after Concorde for the last 30 years and that ATC's job without Concorde was 'changed for ever'. Roger Woodcock thanked ATC for its help and added how much the crews appreciated returning home to the professionalism of the UK's ATC.

On duty in the control tower was Ivor Sims. In 1976, Ivor gave take-off clearance to the very first commercial BA Concorde flight to Bahrain. Today he would be clearing each Concorde to land.

On the airport roads, vehicles cluttered the verges and central reservations to marvel at the spectacle. Concorde had caught the imagination of millions and the BAA was not going to interrupt their homage. When Concorde was new the traffic jams were in protest, now they were in praise.

Back at Heathrow, just after 4.00pm, approaching from the east and having flown from Edinburgh, the first of the three Concordes could be seen. With an attitude of 11 degrees and undercarriage extended, G-BOAE was coming home. For a moment that familiar noise (never an irritant to the devotees of Concorde) reverberated. Too soon it would be confined to the annals of historic recordings. The specially erected grandstand on the north side was now packed with invitees and fans waving their Union flags.

Glimpsed in the sky behind the Edinburgh flight, flew Concorde G-BOAF returning from the Bay of Biscay. Raymond Baxter, an interviewee in the BBC commentary box, purred with pleasure as he caught sight of not one but two Concordes on final approach. The BBC had just aired a clip of him commentating in 1969 on the first ever take-off of Concorde 001. In those days it was hoped that two or three hundred Concordes would be built, making the sight of two Concordes on 'finals' a matter of routine indifference. Then, for a moment in the afternoon sky, three Concordes were discernible on finals.

Prior to the first touchdown a silence descended on the crowd. The gusting north-westerly wind nearly caused Captain Brodie in 'AE' to 'go-around'. With a

Opposite: Concorde 'AF's' final touchdown on runway 27 at Filton. There was a gusty south-westerly wind, but the landing was perfect. (Peter Duckworth, World Airline Photography Ltd)

Overleaf: The Clifton Suspension Bridge was completed in 1864, 98 years before the 'Concorde Treaty' was signed. On 26 November 2003 the crowds turned out in their thousands to welcome another of Bristol's engineering marvels to her birthplace on her final flight to Filton. G-BOAF, Concorde 216, was at 2,000ft, the photographer, Lewis Whyld, at 3,000ft – in a helicopter. (SWNS)

Captain Les Brodie in G-BOAF greets the well-wishers who had come to welcome Concorde back to her birthplace at Filton. He had just completed the final Concorde landing. (Gordon Roxburgh)

sudden loss of headwind and with only feet under the aircraft, Brodie eased back on the control column to avoid arriving too firmly and brushed the runway, before smoothly touching down and lowering the nose wheel – a neatly handled cross wind landing complicated by wind shear. Two minutes and 21 seconds later Paul Douglas landed smoothly in 'AF', decelerated and cancelled reverse thrust before vacating the runway and following 'AE'.

The cameras now turned towards the third and final aircraft, G-BOAG, Flight BA002 from New York (call sign 'Speedbird Concorde zero zero two') approaching from the east. The studio discussion at that moment concerned the nose and visor and what it did for visibility from the flight deck. The subsequent question 'Why weren't more Concordes built?' remained unanswered for, just at that moment, a picture of 'AG' flying away from the camera caused John Sopel to say, 'I want to pause us slightly now, as this is going to be the final Concorde that people are going to see coming in to land at Heathrow. This is it. This is the end of the era.'

Those watching beside the tarmac or on television marvelled at this aeronautical art form – this wing shape capable of spanning over 1,100 knots without the assistance of flaps and slates. As 'AG' slowed for the

approach, the angle of attack increased to generate more vortex lift. Now the extra drag demanded an increase of power; to restore vision ahead the nose was lowered. As the runway approached, the gear was extended, 'down, four greens,' confirmed the crew. This finely honed routine was being played out, with passengers into Heathrow, for the last time. Would that the world had loved the craft at the beginning. 'One hundred feet,' called the engineer, 'fifty,' the autothrottles were disconnected. 'Forty, thirty, twenty, fifteen,' the throttles were closed. The tendency of the slender delta to pitch down (reduce its angle with the horizontal) due to thrust reduction and ground effect was countered with 'up' elevon reducing the rate of descent. The wind was from the right so the drift was removed with left rudder while the wings were kept level with elevon. Bannister's touchdown was perfect; but the applause was as much for 27 years of Concorde as for the landing.

All three Concordes vacated 27R and taxied along the 'outer' taxiway until they were facing east in the vicinity of the 'November' parking stands. With the grandstand to their left and ground staff to their

Filton, 26 November 2003. The final celebration of Captains Bannister and Brodie with Union Jacks. Sad though the occasion was, BA had retired Concorde in style and safety – a fitting tribute for a project that absorbed so much effort by so many people.

The Duke of York accepts the 'Tech Log' of G-BOAF from Mike Bannister on behalf of Filton. On the left Warren Hazelby (Senior Flight Engineer), Rod Eddington and Les Brodie applaud the occasion – 26 November 2003. (Airbus)

right, all three aircraft simultaneously performed the Concorde bow – the sedate lowering and raising of the nose and visor.

The Edinburgh and Biscay Concordes then proceeded on a very slow taxi of honour around Heathrow's stands as hundreds of enthusiastic airport employees clapped and waved. Before the Compass Centre there was another bow. Meanwhile 'AG' was linked to a tug with a ground power unit and towed to the engineering base. Here a red carpet welcome was waiting. Captain Mike Bannister and Senior First Officer Jonathan Napier, leaning out of their 'DV' (Direct Vision i.e. openable) windows, waved Union Jacks. Mike had suffered a soaking from a water arch when the chart holder prevented the quick closure of his 'DV' window but his bonhomie was undampened. The guests assembled in a hangar housing G-BOAD cheered the arriving spectacle. As 'AG' drew up alongside 'AC', steps were manoeuvred into position and a dispatcher knocked on the passenger door. Eventually (reluctantly?) the door opened from the inside. Concorde had finally come home, safely delivering the last 100 of 2.5 million passengers.

First to appear from 'AG' was BA Chairman Lord Marshall who descended the steps looking fit and fresh after his Atlantic crossing. He was greeted by his Chief Executive, Rod Eddington clad in a yellow 'dayglo' tabard to comply with airport anti-collision regulations. Next off the plane was Joan Collins. Sir David Frost followed, only to be waylaid by a microphone on the end of a long pole. There was a 'cordon sanitaire' between the red carpet and the media. 'The only aircraft that lets you be in two places at the same time,' he had said of Concorde. Model Christie Brinkley had her photograph taken under the nose of 'AG' while Jodie Kidd chatted with interviewers across the divide. The Duke of Kent, a passenger on the first BA supersonic flight to Bahrain, walked briskly along the red carpet before turning to watch Mike Bannister extol the 'one big family' that had made Concorde work. 'It's a celebration,' Mike added as he moved on to the next microphone.

A few weeks later, on 26 November 2003, I parked my car on the central reservation of the A3044 directly under the departure path of Heathrow's 27R. There had been heavy rain but that had not discouraged the thousands of people congregating to witness the final take-off of this extraordinary flying machine. Would the crowds have turned out to watch the last departure of a B747?

Sadly there would be no Concorde flying on 17 December 2003 just three weeks later for the 100th anniversary of powered flight. On that day Orville Wright had flown a distance two-thirds the length of Concorde, taking the same time as Concorde would take to cover four and a half miles. Yet the aircraft were not entirely dissimilar. One particular feature of the Wrights' inventiveness was applied to Concorde's flying controls; the Flyer's rudder application was mechanically coordinated with lateral control inputs. Using auto-stabilisation in 'electrical' signalling to the flying

Overleaf: G-BOAF being pushed-back into the Brabazon hangar by a tug after the last ever flight of a Concorde. (Mark Wagner/aviation-images.com)

controls Concorde did just the same. Left elevon would apply left rudder and vice-versa.

Just as the Wrights had settled who would become the world's first pilot – by tossing a coin – so it was resolved who would be the last to land Concorde. Paul Douglas flipped the coin, Les Brodie called 'heads' and won. As he put it, 'it was the Queen who decided.' Paul's compensation was to handle, from the right-hand seat, the last Concorde take-off from Heathrow.

'Here she comes,' someone shouted, heads turned, hands clapped and people cheered. Elegant, purposeful, beautiful, businesslike, sculpted, fast – the hyperboles raced around the mind. The undercarriage retracted, the reheats were switched off and Concorde 216 G-BOAF (the last to be built) was on her way.

She flew a standard departure to the west with acceleration and climb. At 6° 45' W the turn to the left was initiated so that 'AF' remained east of 8° W – the Oceanic boundary. Mach 1.7 'reheats off'; FL500 (50,000ft) 'engine flight rating – cruise; transonic check complete,' the well-rehearsed litany was being sung and acted out for the last time. 'Transonic', that speed regime that had cost the lives of so many test pilots, yet was conquered so supremely by Concorde.

Here, on the threshold of space, if you look up you can make out the midnight blue of the cosmos; to the horizon – the earth's curvature; and if you look down you might just be rewarded by the sight of a jumbo jet, wending its weary way.

Too soon the turn to the east was complete. The Deceleration and Descent Checklist was actioned, the throttles retarded to 18 degrees on the quadrant and the fuel transferred forward. Mach 1.00, the pressure sensing instruments, altimeter, airspeed indicator, Machmeter and vertical speed indicator, gave their characteristic kick, proof that the shockwave had passed forward and that Concorde had left the regime she had ruled for a third of a century.

After the Channel Islands, Concorde turned north to Compton before turning west to fly over Filton towards

However stunning Concorde looks beneath her natural working environment – the sky – the vagaries of corrosion are better avoided. At Manchester Concorde G-BOAC is now housed in a hangar. (© Michael Derby)

the coastline. Thereafter, south-west to Weston-Super-Mare before turning to Bristol International (Lulsgate). From there to the Clifton Suspension Bridge, turning north over the Downs then east towards Marshfield in a 'tear-drop' making a wide left turn to intercept, in the vicinity of Tormarton, the extended centre line of runway 27 at Filton.

The flight crew had previously taken a helicopter ride around the Bristol area so that they could locate the various landmarks. On 26 November, the famous photograph of Concorde 216 over the Clifton Suspension Bridge was taken from that same helicopter hovering at 3,000ft, 1,000ft above Concorde. Many devotees on the Downs had found time to stop and stare. Bristolians built this Concorde, some spent the bulk of their careers caring for her; a project that had shrunk the world for 27 years. Someone had had the wisdom to allocate the last of the Concorde line to Filton.

Les Brodie's wife and daughter were watching his landing from the control tower at Filton. The wind was gusting from the south-west (240°/13). The runway is twice the width of the standard runway, 90m versus 45m. The pilot, therefore, is under the illusion of being lower than he is. Result: high flare, long float, deep landing. With a tail wind the float is longer and there is a hump which obscures the proximity of the end of the runway, which then appears very quickly. The author is writing from first-hand experience. Fortunately the carbon fibre brakes were very effective on that occasion. 'Fifty, forty, twenty, fifteen.' It was a perfect landing. Guests of Airbus at Filton, including the Duke of York, applauded the event.

A short ceremony beneath G-BOAF followed, during which Captain Mike Bannister, who had been on the flight deck, handed over the Technical Log to the Duke of York who accepted possession of it on behalf of Filton. The ownership of all the former BA Concordes still rests with BA, thus preventing possessors disfiguring the aircraft by, for instance, painting on their own logo. There is always the hope that one day 'AF' could once more be made airworthy.

G-BOAF was towed to the Brabazon hanger for decommissioning. This involved the removal of all fluids, the battery, inflatable escape slides and anything that could be dangerous to visitors. The facility to power the aircraft with a ground-power unit was deactivated. There are plans at Filton to have a proper display hangar for 216.

By 2005 every Concorde except G-BOAB had arrived at its final place of exhibition and, although decommissioned, it would not be impossible to make some of them fit for flight again. Claude Freeman, BA

Chief Concorde Ground Engineer, estimated that six months' work would be necessary to return one to flight. The Tu-144 flew again after several years of storage. If the Air Intake Control Units have been removed hopefully they are safely stored. It is hard to believe that these early digital computers still represent leading edge technology.

'Three hundred and twenty thousand pounds, gone,' BANG – not the shock wave from a supersonic plane, but the staggering price paid for a Concorde nose cone. The date was 1 December 2003, the location Olympia, London and the occasion, Bonham's auction of Concorde spare parts that raised funds for the charity 'Get Kids Going'. Even a Machmeter fetched £28,000. A similar sale by Air France in Paris had also astonished cognoscenti with items fetching ten times their reserve price.

Was the Concorde project worthwhile? If it had been known on 29 November 1962 that only 14 supersonic airliners would enter service, the answer would surely have been no. At each stage of the project, however, the answer was yes. The 1964 Wilson government feared French reparations and so desisted from cancellation. Tony Benn, former Labour MP for Filton, replaced the final 'e' once and saved Concorde twice. In the 1980s when the Thatcher government wanted to be rid of the Concorde burden, Brian Walpole and Jock Lowe saved it. Each crisis was overcome until the announcement on 10 April 2003 to cease operations. Concorde had served a generation and, what is more, justified herself. Sad though the end was to be, a planned celebration would sweeten the bitter pill; better a Grand Finale than an ill-defined fade out. Well done BA and Air France for having done justice to a remarkable project. The words on a BA poster summed up the end of the (first?) supersonic era: 'How did 27 years go so fast?'

THE AFTERMATH

I finished my training at the College of Air Training Hamble in February 1969. Hamble was owned at that time by Britain's two nationalised airlines BEA and BOAC. The College was proud of its high standards and believed in educating future pilots to a higher grade than the one which was strictly required to pass the airline pilot licence exams. At the start of training 18 months earlier my course had been asked which airline they would like to go to on completion. I had said BOAC since Concorde would not be operated by BEA.

On 3 March 1969, the day after the first flight of Concorde 001, I joined BOAC. First I was trained to navigate. I learnt how to fix the aircraft's position using a bubble sextant to measure the elevation of three stars, then repeating the process every 40 minutes until sunrise or the destination was found. In the autumn of 1969, once a VC10 had been released from the busier summer schedules, I was able to complete my pilot conversion course to the Vickers airliner flying at Shannon on the west coast of Ireland.

On 4 March 1976 I was on another conversion course, this time at Filton near Bristol, where the British Concordes were still being assembled. I remember watching the television news. Brian Trubshaw was about to land 002 at Yeovilton. To give the landing a sense of occasion a 'batman' (someone who gave signals to a pilot landing on an aircraft carrier) was unnecessarily giving guidance to Concorde. He was so close to the runway

that Trubshaw feared for the batman's life. Trubshaw did not want to do a 'go-around' due to lack of fuel and deteriorating visibility. Consternation … as it was the batman suffered some ear damage. Concorde then became a much-admired exhibit at the museum of the Royal Naval Air Service (RNAS) Museum at Yeovilton. Suddenly I was struck by the thought that I was being trained to fly a museum exhibit or at any rate a later version of one.

Concorde appeared in three distinct forms. There were two prototypes, two pre-production and then 16 production aircraft. The first two production aircraft were 201, assembled in France, and 202 in Britain. Flight testing meanwhile continued. At the same time two Concorde test specimens were undergoing static and dynamic tests.

These first two 'production' versions of Concorde may have looked similar to the following 14, but they would have required substantial modification. Peter Baker, who tested 202 to establish Concorde performance, noted that all four engines were slightly different although they produced the same thrust. The maximum fuel load on that Concorde, G-BBDG, was some 4 tonnes less than the subsequent versions.

Opposite: Air France's Concorde F-BVFA was donated to Smithsonian National Air and Space Museum, fulfilling a promise to the museum made in 1989. F-BVFA was the first Concorde to join Air France's fleet and it had been the first to touch down in the United States. (Studio Alexander Snell)

Adrian Meredith's picture of the final landing at Filton captures a moment in aviation history. In the background 'Rolls-Royce' whose predecessor, Bristol Siddeley Engines, launched the Olympus in 1949. (Adrian Meredith, www.aviation-photos.com)

The fuel instrument displays on the engineer's panel were different. Modifying 'DG' to the same status as the other 14, would, according to Baker, have been more expensive than starting a new Concorde from scratch.

'DG' flew from 1974 to 1981. Then she was parked at Filton. Later she was covered in wax to discourage corrosion before being unceremoniously wrapped in a silver bag. There had been talk of dismantling the aircraft and rebuilding it at Heathrow as a restaurant. The author was asked to assess the viability of the scheme. I feared that dismantling the aircraft would cause irreparable damage structurally and cosmetically (which since the Brooklands experience has proved unfounded). As

Concorde was still in service and there was plenty of opportunity to taste Concorde cuisine at Mach 2, I was very doubtful as to the desirability of entertaining guests at Mach 0. In the end 202 remained in her damp-proof silver sack which was rumoured to be causing rather than preventing the corrosion. She was eventually debagged as a source of parts to keep BA's Concordes flying.

One of the items removed as spares was a firewall. An engine had overheated on a Concorde bound for London and caused a titanium fire in the engine nacelle. The firewall between a pair of engines was on the point of being burnt through when the flames, smothered by quantities of titanium ash, were extinguished. This was evidence, too risky to reproduce in a test flight, that the thickness and therefore weight of the firewall had been designed with exactly the right properties. The design engineers were delighted, whatever their thoughts about the origin of the fire.

Another incident required the replacement of a damaged nose cone. Luckily the cone on 'DG' was suitable. Gradually 'DG' lost the look of a sleek, slender-bodied supersonic airliner, and minus its tailfin metamorphosed into an aluminium tube on three stilts. But she rendered one more service to the Concorde fleet. After the '9/11' atrocities, the accessibility of the flight decks of all airliners had to be reduced. 'DG' was used in the designing of a suitably secure flight deck door.

When Concorde services came rather suddenly to a halt in 2003, no one appeared to know where to exhibit the five Air France and eight BA Concordes. Air France had completed its last service at the end of May 2003 and BA at the end of October. Air France lost no time, and within a fortnight their aircraft were off.

On 12 June 2003 Air France Concorde 205 (F-BVFA) was flown to Washington for exhibition at the Steven F Udvar-Hazy Center – a companion part of the Smithsonian Air and Space Museum.

On 14 June 2003 Air France Concorde 213 (F-BTSD) was flown from Paris Charles de Gaulle to Le Bourget. 'SD' is now displayed alongside Concorde 001 at the Le Bourget Air and Space Museum.

On 24 June 2003 Air France Concorde 207 (F-BVFB) was flown from Paris Charles de Gaulle to Karlsruhe Baden-Baden. Here the aircraft was partially dismantled then taken by barge and road to the Auto & Technik Museum Sinsheim, Baden-Wurtemberg, Germany. 'FB' was a gift from Air France to the museum, and in the handover M Jean-Cyril Spinetta, the Chairman of Air France, paid homage to those who died in Concorde Flight 4590 on 25 July 2000. 'FB' is now spectacularly displayed as if in flight in formation with a Tu-144. The interior of the aircraft is accessible to visitors in spite of the pitch attitude.

'AF' in retirement at Filton.

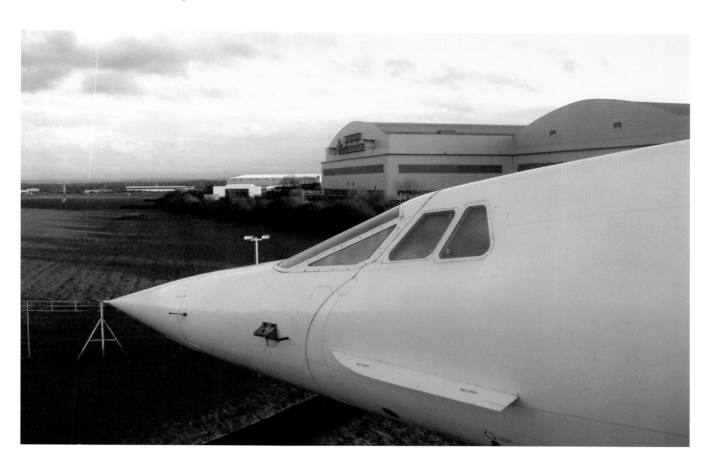

Concorde G-BOAD aboard a barge in New York harbour, en route to the Intrepid Sea, Air & Space Museum where it is displayed, on loan from BA. (US Coast Guard)

On 27 June 2003 Air France Concorde 209 (F-BVFC) was flown from Paris Charles de Gaulle to Toulouse. 30,000 Airbus employees lined the runway to watch the arrival. Concorde had re-established the status of the French aero industry which had been lost due to the Second World War, and now in the 21st century Airbus was vying with Boeing as a leading builder of civil airliners. Disembarking to a rapturous applause was André Turcat, the pilot of the first Concorde flight on 2 March 1969.

The final flights of the BA Concordes had converged on Heathrow on 24 October 2003. With no storage space at the airport, the serviceability of the aircraft running out and the time limit of the pilots' qualification approaching, the flyable Concordes had to go ASAP. At that time only five of the seven were flyable, so those that could fly did. One of those that could not left by barge and the final one (at the time of writing) is still at Heathrow…

Concorde 204 (G-BOAC) arrived at Manchester on 31 October 2003. 'AC' is now on display under cover in a purpose-built hangar. By the time the airline took delivery of 'AC' the airline had changed its name. BOAC and BEA had merged to become British Airways, and G-BOAC's registration lost its obvious significance. Nevertheless this Concorde is regarded as the flagship.

Concorde 214 (G-BOAG) arrived at Seattle on 5 November 2003 – the 45th anniversary of the first sitting of the STAC (Supersonic Transport Aircraft Committee) chaired by Morien Morgan (later Sir). 'AG' is on display at the Museum of Flight. Had Boeing chosen a delta instead of a swing wing design for the US SST, it might not have been cancelled, in which case there would have been many more Concordes and US SSTs flying. Had that been the case, how long would it have taken to make either SST sufficiently quiet?

Concorde 208 (G-BOAB) has remained at Heathrow. It had been hoped that 'AB' could be displayed at Terminal 5, but at the time of writing no plan has been made. There has been a suggestion of dismantling it and reconstructing it at a site near the London Eye on the south bank of the Thames. However this venue may be too close to the exhibit at Brooklands to allow public access – stipulated when 'DG' was taken on at Brooklands.

On 12 September 1981 my wife and I chartered 'AB' for the first of two flights around the Schneider Trophy

Delivered in 1976, G-BOAA was BA's first Concorde. When Concorde was retired G-BOAA was awaiting its post-Gonesse modifications, and it never returned to flight status. It was dismantled and transported via road and sea to East Fortune Airfield, near Edinburgh, where it is now exhibited at the Museum of Flight. (© Gordon Barnes)

F-BTSD was retired to Le Bourget Air and Space Museum, Paris. It is on show alongside the French prototype Concorde 001, F-WTSS. (© Andrew Scorgie)

course of 1931. On 13 September we took 'AG'. On both days we took 100 passengers beyond Land's End at Mach 2, then descended and decelerated eastbound over the English Channel. We turned north to cross the Isle of Wight and join the course near Calshot Spit, thence to Ryde Pier, then left towards West Wittering (east of Portsmouth). Next we flew west along the coast back to Calshot where the 50th anniversary celebrations were taking place. After that we flew the course in the reverse direction.

West Wittering was the former home village of Sir Henry Royce. He famously sketched an engine in the sand when asked in 1928 whether Rolls-Royce would consider building an engine for the 1929 Schneider contest. The Rolls-Royce Merlin and Griffon engines benefitted greatly from the experience gained by Rolls-Royce from the racing ('R') engines. RJ Mitchell designed the victorious Supermarine S6 and S6b of 1929 and 1931 respectively. Then he went on to design the Spitfire. In 1931 George Stainforth took the world airspeed record to 407mph. The High Speed Flight of the RAF had been led by a distant cousin of the author – Harry Orlebar (see page 94). The author says he receives more accolade than is due owing to the unusualness of the name.

On 10 November 2003, Concorde 210 (G-BOAD) arrived at JFK New York. 'AD' was transported by barge to the west side of Manhattan Island to be a feature alongside the aircraft carrier USS *Intrepid*. The left-hand side of 'AD' displayed Singapore Airlines livery for the short period that the route from Bahrain was extended 1979–1980.

Concorde 212 (G-BOAE) arrived at Barbados on 11 November 2003. 'AE' is on display and under cover at Grantley Adams airfield.

In April 2004 Concorde 206 (G-BOAA) was dismantled at Heathrow, taken by road to Isleworth and loaded on a barge. When the neap tides came, clearance under the Thames bridges was assured and 'AA' set sail for Scotland. Reconstructed at the East Fortune Museum, 'AA' is a major attraction.

Concorde 216 (G-BOAF) made the final landing of all the Concordes at Filton on 26 November 2003. The management in charge of exhibiting 'AF' had hoped that if 'DG' was kept, then Filton could show one without furnishings – displaying the innovative engineering which allowed for the expansion and contraction

G-BOAC retired to Manchester Airport, where it is now located under cover in a glass hangar. (Paul Heslop)

'DG' COMES HOME TO BROOKLANDS

1 G-BBDG on a 'jury' undercarriage at Filton, Bristol. BA had 'borrowed' pieces to use as spare parts to keep its existing fleet serviceable. The nosecone had gone, as had a firewall between two of the engines, and the fin. 'DG' had not flown since December 1981. Concorde needed a 2,000m-long runway for landing; although it could stop in a shorter distance. The 1,100m one at Brooklands was too short and in any case due to be closed to make way for the Mercedes-Benz test circuit. A Concorde could only be delivered by road to Brooklands and 'DG' could only be moved by road from Filton. QED?

2 April 2004. The limiting factor for road transport is the width of the load. So 'DG' had to be dismantled into the largest pieces possible consistent with loaded width. Here a wing section arrives at Brooklands.

3 The 'homecoming' of Concorde 202 G-BBDG, 5 May 2004.

4 Toasting the arrival of 'DG' from Filton, 6 May 2004, are from left to right, Peter Baker (Assistant Chief Test Pilot Concorde), Alan Heywood (Flight Test Engineer), Roy Radford (Test Pilot), Alan Smith (Training Pilot) and Johnnie Walker (Test Pilot) – all were BAC (later BAe) personnel. Peter Baker was the pilot of the Canberra chase plane that accompanied 002 on its maiden flight on 9 April 1969. On 24 December 1981 he landed 'DG' at Filton on its final flight.

5 September 2005. 'DG' at Brooklands without the outer wings and still awaiting an undercarriage; but looking more like an SST.

6 'DG' looking brand new and elegant. Between 30,000 and 35,000 people visit Concorde annually.

7 21 June 2006. After some discussion it was decided to use the British Airways livery in which 'DG' originally flew.

8 In order to give a taste of Concorde in flight a member of the Brooklands volunteer staff takes a group of 15 'passengers' at a time on a flight. They receive an introduction in the static airline bus adjacent to 'DG' followed by a walk and talk beneath the aircraft before boarding. Entrance is via the rear baggage hold, minding one's head on the low door. This picture shows the view of the rear cabin with seats removed and various items on display. Thence the tour takes the visitors to the front cabin where they are seated (no seat belts) in order to view an abbreviated version of a transatlantic flight on a video complete with the cabin information displays (speed, Mach number etc), which is presented by Captain Mike Bannister, former Concorde Chief Pilot.

9 Inside the marquee, a view of 'DG' from the flight deck of the Concorde simulator. The simulator could have been kept in running order at Filton, but to make room for other projects it was dismantled with no expectation of reuse. However with great skill, determination and improvisation, an electronics expert – Gordon Roxburgh – is, at the time of writing (October 2010), continuing to improve the simulator's performance. Members of the public can, under the tuition of former Concorde pilots and for a fee, experience flight at Mach 2 including a view of the earth from 60,000ft on a 165-degree panoramic screen.

Two supersonic airliners are now displayed on the roof of the Sinsheim Auto & Technik Museum, Concorde F-BVFB and Tu-144 77112. (Stuart Pearce)

Above and right: In Barbados, Concorde G-BOAE (212) is exhibited in a specially built hangar. A *Son et Lumière* presentation is available, with reduced 'son' on take-off. There are two simulators with which visitors can test their skill at supersonic flight. (Christopher Begley)

experienced during a supersonic flight – and the other fully furnished as the passengers would have found her.

Concorde 202 (G-BBDG) was offered to the Brooklands Museum Trust in October 2003. The aircraft would have to be cut up into sufficiently small pieces to satisfy road regulations. The illustrations show the rest of the story. Now the visitor, seated in the front cabin of 'DG', can view a video of a flight narrated and acted out by Captain Mike Bannister from the flight deck of 'DG' itself.

If the visitor is still wishing to get to grips with the Concorde technology, he can for further fee, fly the simulator under the instruction of a former Concorde pilot. The flight deck of the simulator was disconnected from its computer at Filton and delivered to Brooklands with little hope of it ever working again. Thanks to the electronic wizardry of Gordon Roxburgh, who has most

generously given hours of his time, the simulator is coming back into operation. There is no motion but the visual has been so improved that the impression convinces the viewer he is airborne over a real world.

One more story in the aftermath has been the legal case by the French judiciary over the negligence of anyone who might have caused the crash. There is little to gain, I would suggest, by pressing home the case. Concorde is not flying; the designer in question, if he were to design again, would not repeat the same mistakes, even if Concorde's weaknesses should be considered 'mistakes' in the first place.

THE FUTURE

Predicting the future is not easy. Inventions and discoveries are continually made, sometimes their appearance upsets the status quo; occasionally established beliefs are destroyed. Early in the 17th century Galileo, using just two lenses, constructed a telescope. He observed Jupiter's moons and worked out that they were in orbit around that planet. From this he deduced that the sun, not the earth, was at the centre of the solar system. He had reduced the status of God's earth and found himself in serious trouble from the Church. However it broke the logjam caused by the insistence that the geocentric view (that Earth was at the centre of the universe) was correct. No one in 1560 had predicted that by 1610 there was going to be a telescope and that it was about to spearhead the scientific revolution. Are we now more capable at making accurate predictions than mankind was in the 17th century? Will some development occur which will overturn our established beliefs?

We make predictions based on sets of assumptions. In 1962 the Concorde project was justified by several assumptions: passengers would pay extra for halved flight times; the price of fuel would come down; there would be no serious inflation; the engine noise, equalling that of the aircraft of 1962, would be acceptable; the US airlines would buy the SST; the

sonic boom would be tolerated over populated areas; and that most long-haul passenger flights would be supersonic. In 1976 when Concorde services began, fuel prices were as high as ever due to the Arab–Israeli war, inflation had made the cost of Concorde appear to be outrageous, the big fan engines of the new Boeing 747 were very much quieter than those of the 1962 airliners, the US had taken an anti-Concorde stance and the sonic boom was intolerable. The first assumption (that passengers would pay extra for halved journey times) proved to be correct and it saved the first 14 Concordes.

Since the late 1980s it has been constantly predicted that man-made CO_2 emissions will cause the climate to change. Evidence to support this includes warmer winters, the retreat of glaciers, the melting of polar ice and the analysis of ancient air bubbles trapped in ice. That the UK suffered a cold winter in 2009–2010 was said to be a variation in the weather rather than a change in the climate. Whether the prediction is correct or not is academic. The belief is that CO_2 emissions should be reduced and that includes emissions from aircraft.

Airliners are responsible for only 2% of the total quantity of CO_2 emitted into the atmosphere. Nevertheless ways of reducing this are being researched. The Boeing 787 Dreamliner appeared at the 2010 Farnborough Air Show. Built of carbon composites rather than aluminium alloy it has the same passenger capacity as, but is much lighter than, its equivalent – the Boeing 767. As a result it uses 20% less fuel.

Above: Ideally a successor to Concorde would need a variable cycle engine that could switch from being having a high bypass for subsonic flight, as seen attached to a wing on the right of this picture and a turbojet for supersonic flight, like the Rolls-Royce Olympus seen on Concorde.

Below: Quiet Spike was a joint project by NASA and Gulfstream to cut the noise signature of a supersonic aircraft at ground level. The 24ft composite spike, patented by Gulfstream, designed to divide the bow shockwave into a series of less-intense pressure waves, reducing the sonic boom. Gulfstream's interest stemmed from its plans to develop a supersonic business jet jointly with the Russian aerospace firm Sukhoi. Micro waves focused on a point ahead of the SSBJ might achieve the same effect. (NASA/Carla Thomas)

Engine technology could give even greater savings. For improved propulsive efficiency the velocity of the propelling gas (the exhaust velocity) must be as close to the forward speed of the aircraft as possible. Early jet engines consisted of an intake, a compressor, combustion chambers, a turbine and an exhaust (propelling) nozzle. All of the air entering the intake exited via the exhaust, and the faster the exhaust the greater the thrust. A larger mass flow of air at a given velocity also increases the thrust. Later an engine with a 'fan' driven by an extra turbine via a coaxial shaft was designed (coaxial means that one shaft is within another). Behind the fan the airflow splits; some enters the engine core and the rest bypasses the core. The bypass ratio is the mass of air bypassing the engine core divided by the mass that goes through the core. At subsonic speeds a fan engine is quieter and more fuel efficient than a turbojet engine. The first set of 'big fan' engines in the 1970s had a bypass ratio of about 5; the latest Rolls-Royce Trent engine exceeds 9.

The Aerion SSBJ is currently the supersonic business jet project most actively under development, and should be capable of cruising at up to Mach 1.6 over the oceans and other uninhabited areas. As conceived, the baseline Aerion aircraft has a relatively low sonic boom, with an initial overpressure of about 0.8psf. This is less than the boom of many supersonic fighters and much less than the Concorde. For flights over the United States, where aircraft must stay below Mach 1.0, the Aerion will cruise at Mach 0.98. In other populated parts of the world the regulations require merely that a sonic boom does not reach the ground. There, the design of the Aerion will allow it to cruise as fast as Mach 1.1 without creating a sonic boom on the ground: the path of the shockwave, which is at right angles to the wave front, is refracted away from the surface due to the difference in the temperature of the air around the aircraft and that of the air at the surface. The greater the difference, then the greater will be the refraction. The refraction of the path of the shock wave is analogous to the refraction of light causing a mirage. This effect is greater when the aircraft is flying in a strong headwind compared to the wind on the surface, but reversed in a tailwind. Mach 1.1 would not have been an economic speed for Concorde. (Courtesy Aerion Corp)

It also has three coaxial shafts which improve the air compression. This 'triple spool' technology was first applied to the RB 211 in the late 1960s.

A high bypass fan engine would have too low an exhaust velocity to give thrust at supersonic speeds. Concorde was fitted with a turbojet engine. The Olympus 593 could give plenty of thrust at slow speed (less than 250 knots) but it made too much noise and was inefficient. In contrast at Mach 2, it was the world's most efficient engine. Concorde went ten times further on a gallon of fuel at the latter end of the supersonic cruise than when on final approach. There was a very small way of increasing the mass flow on the Olympus 593. The low pressure 'spool' (shaft turning the seven low pressure stages) could be made to rotate faster with respect to the high pressure 'spool' (shaft turning the seven high pressure stages) by increasing the primary nozzle area. After take-off the engines would switch to 'fly over' which helped a little to reduce engine noise. In 1962 an engine with the kind of variable cycle (variable bypass ratio) that was needed was not available.

Previous page: Sir Richard Branson, with his eye on the potential space tourism market, is having built (October 2010) a suborbital two-stage launch system. When ready it will take six fare-paying space tourists for a ride into space and back to where they started. During the flight they will achieve Mach 3, experience six minutes of zero gravity, see the earth from an altitude of 110km (six times higher than Concorde) and return three and a half hours later to the spaceport in the Mojave Desert, USA. Tickets will cost about $200,000 each. If developed it could deliver passengers to other destinations. If so it would become the world's first suborbital transport. Reduction of the ticket price will be a major challenge. The spacecraft itself is the craft slung between the two outer fuselages. (Virgin Galactic/Mark Greenberg)

In 1976 there was a proposal to change the design for Concorde 217 onwards. The 'B' Wing would have had a greater wing area with a moveable leading edge – improving subsonic performance and reducing noise – extra fuel capacity, and on the engine an extra compressor stage with a different turbine, no reheat, an air bleed underneath and acoustic lining to reduce noise. With four Concordes unsold, the will and the finance for this radical improvement were not there; production ceased with Concorde 216. Had the 'B' Wing flown, the future would have been very different. For one thing would BA have been happy flying the Concorde 'A' Wing with a competitor flying the 'B' Wing? Nevertheless environmental considerations would still have been forcing designers to think of yet higher-bypass engines to save CO_2 emission on the latest subsonic types.

To return to the non-variable cycle engine of the future. The limit in bypass ratio is reached when the fuel it saves is less than the extra fuel needed to carry the heavier engines. If the higher bypass could be achieved without the extra weight of a nacelle, it would be worth doing. The 'free rotor' or 'unducted fan' engine is designed to fulfil that role. The A400M has four Europrop TP400-D6 engines each with an eight-bladed variable pitch propeller. This power plant gives the idea of what a Free Rotor might look like, but with more blades and no variable pitch. The cruise speed of the A400M is from Mach 0.68 to Mach 0.72. Compared to the A380 at Mach 0.85, this would add over an hour to the flight time over a 3,000nm distance.

Concorde used to 'trip the light fantastic'. Westbound the aircraft arrived before it had left, although only in terms of local time. Eastbound the passengers lost weight, 'thereby justifying the superb cuisine', admittedly returning to normal weight once they had stopped. The eastbound effect is due to the flight being in the same direction as the earth's spin. After launch all heavy weight loads bound for space, head to the east. The spin of the earth at the equator gives the launcher about 1,000mph at no extra cost. But so far an aircraft has not taken off from a runway and flown into space.

Alan Bond, founder and Director of Reaction Engines Ltd is working at the Science Centre Culham near Oxford. He and his team have every intention of building an engine capable of flying a 'space plane' into low earth orbit in a single stage. Mastery of such technology will allow the development of a 300-seat hypersonic (faster than Mach 4) transport capable of flying halfway round the world.

Skylon is the name of the proposed 'space plane' and Sabre the name of the engine. The Sabre engine in air-breathing mode will give thrust from a standing start up to Mach 5, beyond which it switches to rocket mode. By Mach 5 the temperature of the airflow entering the engine rises to 1,000°C as it is slowed inside the intake. Adding more heat to obtain thrust would melt the combustion chamber and waste the heat energy in the newly compressed air. So during the acceleration a precooler in the intake of the Sabre engine will, at the extreme, cool the air down from 1,000°C to -140°C. 2,000km of 1mm bore tubing made of an Inconel alloy will be circulated with cooled helium, heating the helium up enough to turn a turbine

The holy grail of space flight, a craft that can take off and fly into low earth orbit and then return to earth without shedding fuel tanks or rocket boosters – in other words a 'Single Stage To Orbit' (SSTO) craft. These concept illustrations depict Reaction Engines' proposed Skylon.

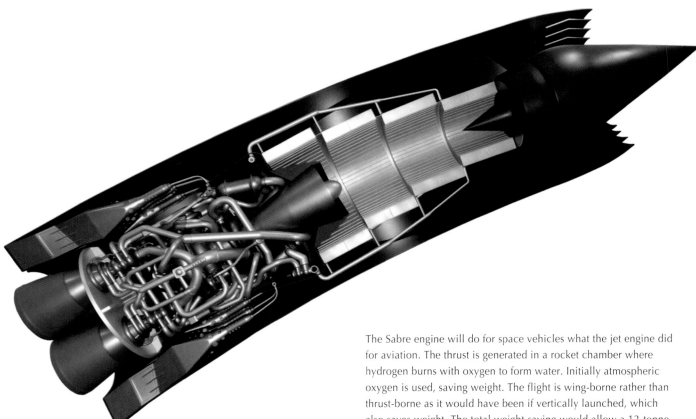

The Sabre engine will do for space vehicles what the jet engine did for aviation. The thrust is generated in a rocket chamber where hydrogen burns with oxygen to form water. Initially atmospheric oxygen is used, saving weight. The flight is wing-borne rather than thrust-borne as it would have been if vertically launched, which also saves weight. The total weight saving would allow a 12-tonne payload to be carried into low earth orbit.

Air enters the intake and is slowed down, which compresses it and causes it to heat up. The flow of hot air is cooled as it passes through the precooler. At Mach 5 the airflow is cooled from 1,000°C to -140°C. The heat energy is passed to the helium. Next the hot helium powers a turbine. The turbine turns a compressor which compresses the cold airstream and in doing so heats it up to 450°C. Finally the air enters the rocket chamber where it is burnt with the hydrogen to give thrust. Meanwhile the helium is cooled in a heat exchanger itself cooled by the liquid hydrogen fuel. Faster than Mach 5, oxygen carried in the spacecraft is used instead instead of air. (Courtesy Reaction Engines)

which drives a compressor which will further compress the air which is then fed into the combustion chamber. Finally the helium is cooled again in a heat exchanger. Here the coolant is the liquid hydrogen fuel on its way to the combustion chamber. Once the speed exceeds Mach 5 the combustion chamber is fed with oxygen from the liquid oxygen tank with the intake source being shut off.

There, in a nutshell, is the process. Omitted are some of the finer points such as disposal of excess hydrogen if it were required to lower the temperature of the helium. The next stage is to prove beyond doubt that the precooler is viable. By autumn 2011 a test rig using a Rolls-Royce Viper engine will, it is hoped, have proved the soundness of this extraordinary work of plumbing.

Once it works, then another project comes a step closer. Long-Term Advanced Propulsion Concepts and Technologies (LAPCAT) is name of this study. The A2 version is now under scrutiny. It will be powered by four

Scimitar engines. Instead of performing during only ten minutes of accelerating flight, the Scimitar precooler will have to work continually for four or five hours, in which time it will have reached the Antipodes. Then after refuelling with liquid hydrogen, return to its point of departure.

The Scimitar will be a variable cycle engine. Noise on take-off will not be the problem that it was for Concorde. Subsonic range will be the same as that when hypersonic. The emissions will not include CO_2.

THE FLIGHT OF THE LAPCAT A2

1 Windowless, loaded and refuelled, the A2 'Lapcat' (Long-Term Advanced Propulsion Concepts and Technologies) is ready for departure. So far this aircraft is only a study of what may be possible. Once hydrogen fuel becomes readily available without causing extra CO_2 emissions such a 'study' may become a 'reality'.

2 Complicated noise abatement techniques like those flown on Concorde would be unnecessary, as the Scimitar engine working in 'high bypass' mode would not be noisy.

3 Due to the variable cycle Scimitar engine, the range of the A2 would be unaffected by speed – unlike Concorde which lost 25% range when subsonic.

4 Once supersonic, the big fan would no longer be powered. With no bypass ratio very efficient thrust is given for supersonic flight from the high velocity exhaust. As the speed approaches Mach 5 the precoolers in the intake will be dropping the air temperature down from 1,000°C to -140°C.

5 At Mach 5 the temperature rise at the 'stagnation point' (where the airflow comes to rest, eg on the tip of the nose) is about 1,000°C.

Elsewhere it is cooler. Nevertheless the leading edges would glow a dull red heat visible at night. The maximum stagnation temperature on Concorde was 127°C.

(All images courtesy Reaction Engines)

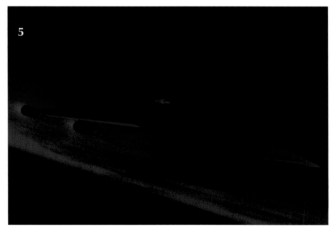

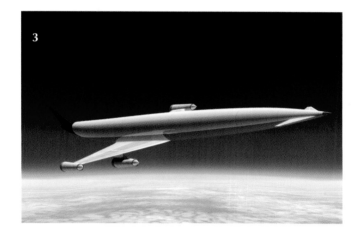

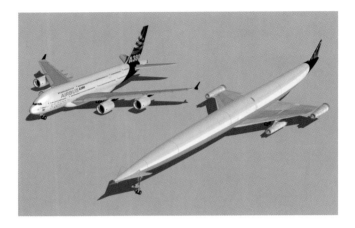

Longer and narrower than the A380, the A2 is expected to have a take-off mass of 400,000kg for the A380's 569,000kg, carry 300 passengers against 555+ for the A380, take 198,000kg fuel (liquid hydrogen) versus 260,000kg (kerosene), have enough range to reach the Antipodes – 20,000km, the A380 having about 15,000km range, the A2 would cruise between Mach 4 and Mach 5 climbing to 90,000ft (Concorde 60,000ft) but the A380 at Mach 0.85 has a 'service ceiling' of 43,000ft (when light enough).

Care is being taken that nitrous oxides are not emitted to excess either. The diameter of the pod will be sufficient to mount a 'big fan' with a 'hub' turbine. The ducts to the rear of the big fan could be fitted with reheat. The fan would be unpowered above Mach 2.5.

The A2 craft itself will be windowless. All exterior views will be shown on flat screens, maybe even in 3-D if Moore's law continues to operate. Complete control from the ground would be standard practice; on board there would be an airline 'technician' who could tell the ground what to do with the progress of the flight even if he could not take over control himself. The A2 could be 'flown' by one pilot providing there was a facility on the ground capable of taking over control. Nevertheless having two 'hostages' on board with them would make the passengers feel even more secure.

LAPCAT A2 Design Mission

Brussels to Sydney (via North Pole and Bering Straits to avoid supersonic overflight of Eurasian land mass)

Distance:	18,700km
Flight time:	4.6 hours (under realistic air traffic control conditions)
Reserve range:	5,000km @ Mach 0.9
Payload	300 passengers (plus baggage)

Vehicle parameters

Gross take-off mass:	400 tonnes
Fuel mass:	198 tonnes (liquid hydrogen)
Fuselage length:	139m
Fuselage diameter:	7.5m
Wingspan:	41m
Wing area:	900m²

Performance

Mach 0.9 cruise Lift/Drag 11.0 (5.9km altitude) SFC 96.0 kN/kg/s

Mach 5.0 cruise Lift/Drag 5.9 (25–28km altitude) SFC 40.9 kN/kg/s

Take-off sideline noise @450m 101.9 dBA

Costs

Estimated 13-year development programme @ €22,600 million (2006 prices)

Vehicle sale price €639 million (assuming production run of 100 vehicles)

One-way ticket price (Brussels–Sydney) €3940

A great opportunity for British engineering is just appearing. In 2010, the US Space Shuttle is close to the end of its life. Rocket-propelled vertical take-off using multiple stages remains hideously expensive. Assuming there is a need to exploit space or at least the objects in it and at the same time satisfy mankind's instinct for exploration, now is the moment to promote the development of the precooler technology which is the key to future.

Let us never forget that in spite of all its setbacks, Concorde proved to be a great inspiration to many. Even in 'museum mode' it attracts enormous adulation, and at Brooklands the Concorde simulator has been re-established, admittedly not in its original form, where visitors can 'fly' under the tutelage of an expert. Concorde was overwhelmingly voted Icon of the 20th Century. Vive le Concorde.

APPENDICES

Max take-off weight	185,070kg (408,000lb)
Max landing weight	111,130kg (245,000lb)
Max weight without fuel	92,080kg (203,000lb)
Max payload (approx)	13,150kg (29,000lb)
Max number of passengers	128 (but BA Concordes were fitted with 100 passenger seats)
Max fuel at specific gravity 0.80 limited by volume	95,680kg (26,400 gallons)

BREGUET RANGE EQUATION

The maximum range is constrained by several factors including payload, reserve fuel required on arrival, weather conditions and the amount of the route that can be flown supersonically. A typical maximum range on Concorde with 100 passengers plus baggage (10,000kg total payload) and normal fuel reserves was 3,500nm. The route between Bahrain and Singapore was flown in the colder air found over the tropics; this improved the efficiency of the engines allowing a range of 3,720nm to be flown regularly. Although the payload limit on this sector was 7,500kg (75 passengers) it could usually be increased to close on 100 passengers when conditions were favourable. The maximum recorded range on Concorde under very favourable conditions, and therefore difficult to plan in advance, was between Washington and Nice, a distance of 3,965nm flown on 11 September 1984 with 54 passengers in G-BOAB, taking four hours and seven minutes.

The range (R) in nautical miles (no wind) of an aircraft can be expressed as:

$$R = \frac{V \times L/D}{SFC} \times \log_e (w_1/w_2)$$

Where V is the cruising speed in knots; L/D the lift to drag ratio; SFC is the specific fuel consumption of the engine – pounds of fuel per hour per pound of engine thrust; (w_1/w_2) is the ratio of weights at the start and end of the sample of range to be investigated. The natural logarithm of the final term is a measure of the structural efficiency of the aircraft – the amount of fuel available for aircraft weight (including payload). For Concorde travelling at Mach 2, V was circa 1120knots; L/D was 7.5; SFC was 1.18 (each engine consuming fuel at just over 11,000lbs per hour would give, during the cruise, about 9,300lbs of thrust). If the two weights were 165 (w_1) and 110 (w_2) tonnes (t) respectively, Concorde could have travelled just under 2,900nm on 55 tonnes of fuel. Should the weights have been 155 tonnes and 100 tonnes respectively, then the range would have improved to 3,120nm. In practice over 90 tonnes of fuel was loaded including approximately 15 tonnes of 'spare' fuel for diversion etc, and much of the fuel was burnt during the 'off design' period of the flight. However, the equation can be applied to each segment of the flight.

CONCORDE – LEADING DIMENSIONS

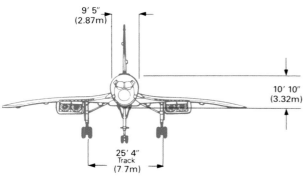

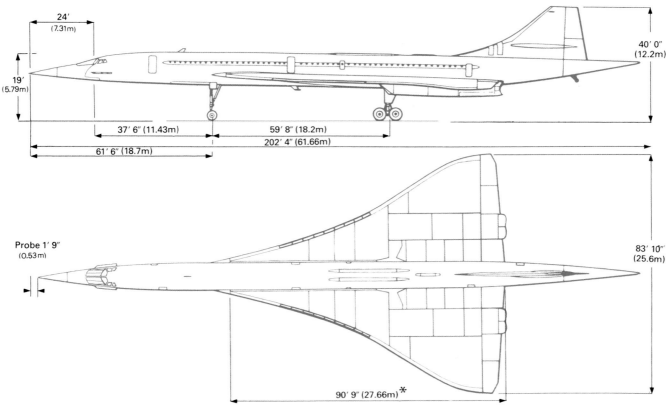

POWERPLANT

Four Rolls-Royce/SNECMA Olympus 593 Mark 610 turbojet engines. Maximum thrust at take-off with afterburner (reheat) contributing about 20% to the total 38,050lb.

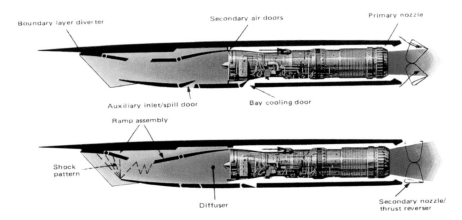

Left: At take-off 95% of the thrust is delivered by the engine and 5% by the nozzle and intake assemblies. During supersonic cruise only 50% of the thrust is delivered by the engine; the remainder is delivered, equally shared, by the variable ramp assembly and the convergent/divergent nozzle system. The purpose of the diffuser is to slow, and therefore compress, the air, which is now subsonic, still further before it enters the engine face. Note: A convergent passage slows and compresses a supersonic airflow, while a divergent passage slows and compresses a subsonic airflow.

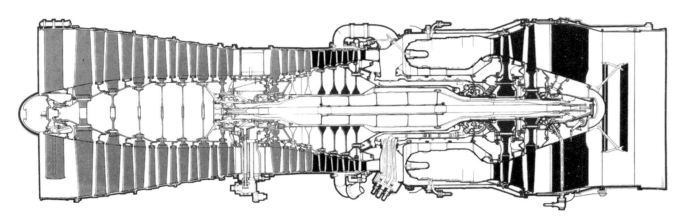

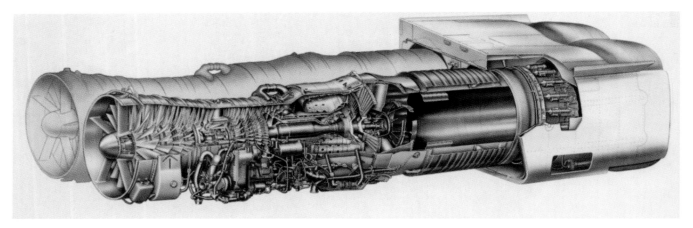

Above: The Olympus engine is a two-spool engine. The inner, or low-pressure (lp) shaft revolves within the outer, or high-pressure (hp) shaft. There are 14 compressor stages, seven on each shaft, driven by their respective turbines. As the air approaches the combustion chambers, during the supersonic cruise, it becomes very hot due to its compression (over 80:1 total). hence the need to construct the final four compression stages from a nickel-based alloy – usually reserved for the turbine area. (The darker shades, green in the cutaway engine, show the more heat-resistant alloys.) To the rear of the turbines is the reheat assembly. The amount of fuel burnt in the combustion chamber controls the rpm of the hp shaft, whilst varying the area of the primary nozzle (see top) controls, at a given rpm of the hp shaft, the rpm of the low pressure shaft.

THE FLIGHT ENVELOPE

Concorde had to be flown within the slightly shaded area shown on the graph. The maximum operating indicated airspeed limit (V_{MO}) varied not only with altitude, but also, to a small extent, with Concorde's weight. Above 51,000ft the maximum operating Mach limit (M_{MO}) overruled V_{MO}; by 60,000ft M_{MO} (Mach 2.04) corresponded to 440 knots indicated airspeed. The lowest authorised speed limit below 15,000ft is not shown on this graph since it was related either to weight on take-off from a particular runway, or to landing weight.

Although M_{MO} is Mach 2.04, Concorde generally cruised at Mach 2. There were, however, further constraints to the speed. One was derived from the position of the centre of gravity; at the furthest forward position of the centre of gravity, the top Mach number limit was Mach 0.8; during acceleration the centre of gravity was moved progressively rearward and at its most rearward point the Mach number ranged from Mach 1.55 to over Mach 2.04; however, the top speed was then limited by M_{MO}, V_{MO} or T_{MO}. T_{MO} was the maximum operating temperature limit, representing another constraint to the speed; the maximum temperature at the nose must not exceed 127°C (261°F). In practice this did not become a limiting factor to Concorde until the outside air temperature was warmer than –50°C (–58°F), at altitudes above 48,000ft.

Next to the altitude, in thousands of feet, is shown the outside air temperature corresponding to the International Standard Atmosphere (ISA). Next to this is shown the speed of sound for that given temperature. The true outside air temperature is usually at variance from the 'standard' by a few degrees either way over the North Atlantic, so the figure for the speed of sound must be regarded as approximate. Over the equator, where the tropopause is higher, temperatures between 50,000 and 60,000ft are of the order of ISA –20°C (i.e. –77°C). At this temperature the speed of sound is 545 knots (627mph).

The formula for the speed of sound (a) is

$a = constant \times \sqrt{absolute\ temperature}$

assuming temperature in °K and speed in knots, the constant = 38.89

For an outside air temperature of –57°C (216°K)

$a = 38.89 \times \sqrt{216}$
$= 571\ knots$

The formula for giving the approximate rise in temperature (ΔT)°C for a given speed is

$$\Delta T = \left(\frac{speed\ in\ mph}{100}\right)^2$$

Thus at Mach 2 when the outside air temperature is –57°C the temperature rise is

$$\left(\frac{2 \times 675.5}{100}\right)^2$$

$$\simeq 173°C$$

Under these conditions the total temperature is $173° - 57° = 116°C$.

Note At the speeds associated with a Space Shuttle (or Skylon) re-entering the earth's atmosphere this formula is not applicable.

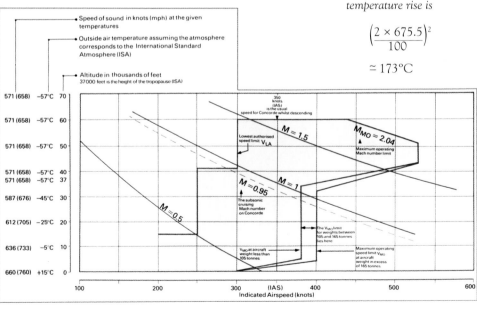

ANATOMY OF CONCORDE

1. Pitot head
2. Radome
3. Nose drooped position (17.5° down)
4. Weather radar scanner
5. Radar equipment module
6. Radome withdrawal rails
7. Radar mounting bulkhead
8. Visor operating hydraulic jack
9. Pitot head port and starboard
10. Visor retracting link
11. Retracting visor
12. Drooping nose operating dual screw jacks
13. Visor rails
14. Incidence vane
15. Front pressure bulkhead
16. Droop nose guide rails
17. Forward fuselage strake
18. Droop nose hinge point
19. Rudder pedals
20. Captain's seat, First Officer's to starboard
21. Instrument panel, analogue
22. Internal windscreen panels
23. Overhead systems switch panel
24. Flight Engineer's station
25. Swivelling seat
26. Direct vision opening side window panel
27. Observer's seat
28. Circuit breaker panels
29. Avionics equipment racks, port and starboard

30. Starboard service door/emergency exit
31. Forward galley units port and starboard
32. Main entry door
33. Air exhaust vents, equipment cooling
34. Life raft stowage
35. Forward toilet compartment
36. Wardrobes, port and starboard
37. VHF antennae
38. Four-abreast passenger seating
39. Cabin window panels
40. Nose undercarriage wheel bay
41. Floor support structure above nosewheel bay
42. Nosewheel leg strut
43. Twin nosewheels, forward retracting
44. Spray suppressor
45. Nosewheel steering jacks
46. Telescopic rear strut
47. Nosewheel leg pivot mounting
48. Hydraulic retraction jack (2)
49. Retractable landing/taxiing light
50. Ventral baggage door
51. Underfloor baggage hold
52. Forward passenger cabin, 40 seats in BA 100-seater layout
53. Overhead light hand baggage rack
54. Cabin air duct
55. Passenger service units
56. Toilet compartments, port and starboard
57. Mid-cabin doors, port and starboard
58. Cabin attendant's folding seat
59. Stowage lockers, port and starboard
60. Fuselage skin panelling
61. Rear 60-seat passenger cabin
62. Cabin floor panels with continuous seat rails

63. Fuselage fuel tank roof panels
64. Conditioned air delivery ducting to forward cabin and cockpit
65. Fuselage conventional frame and stringer structure
66. Wing spar attachment double main frames
67. Starboard main undercarriage stowed position
68. Undercarriage bay central keel structure
69. Port main undercarriage wheel bay
70. Pressure floor above wheel bay
71. Rear cabin conditioned air delivery ducting
72. Cabin wall insulation
73. Cabin floor carried on links above stressed tank roof
74. Foot level cabin ventilating air duct
75. Cabin wall trim panelling
76. Dual ADF antenna fairings
77. Starboard main undercarriage pivot mounting
78. Inboard wing skin with tank access panels
79. Leading edge ventral Spraymat de-icing
80. Starboard wing main integral fuel tanks
81. Outer wing panel joint
82. Fuel/hydraulic fluid air heat exchanger
83. Fuel/air heat exchanger
84. Engine fire suppression bottles
85. Starboard wing fuel feed tank
86. Fuel-cooled engine bleed heat exchangers

87. Conditioning system cold air units
88. Engine bay heat shield
89. Outer wing panel integral fuel tank
90. Tank skin with access panels
91. Fuel pump in ventral fairing
92. Elevon hydraulic actuators in ventral fairings, fly-by-wire control system, electrically signalled
93. Mechanical trim and control back-up linkage
94. Dual outboard elevons
95. Starboard engine primary exhaust nozzle shroud
96. Combined secondary nozzles and reverser buckets
97. Starboard inboard elevon
98. Rear service door/emergency exit, port and starboard
99. Cabin rear bulkhead with stowage lockers
100. Rear galley units
101. Rear avionics equipment bays, port and starboard
102. Oxygen bottles
103. HF notch antennae
104. Fin leading edge structure
105. Multi-spar and light horizontal rib fin structure
106. Lower rudder hydraulic actuator
107. Upper rudder hydraulic actuator in starboard fairing

108. VOR antenna
109. Upper rudder segment
110. Rudder aluminium honeycomb core structure
111. Lower rudder segment
112. Extended tailcone fairing
113. Tail navigation light
114. Fuel jettison
115. Flight data recorder
116. Nitrogen bottle
117. Retractable tail bumper
118. Fin rear spar and tail bumper support bulkhead
119. Fin spar attachment joints
120. Rear fuel transfer tank
121. Fin spar support structure
122. Rear pressure bulkhead
123. Starboard side baggage door
124. Rear baggage compartment
125. Wing trailing edge root fairing
126. Port inboard elevon
127. Machined elevon hinge rib
128. Inboard elevon actuactor in ventral fairing
129. Port wing rear main and feed integral fuel tanks
130. Machined wing spars

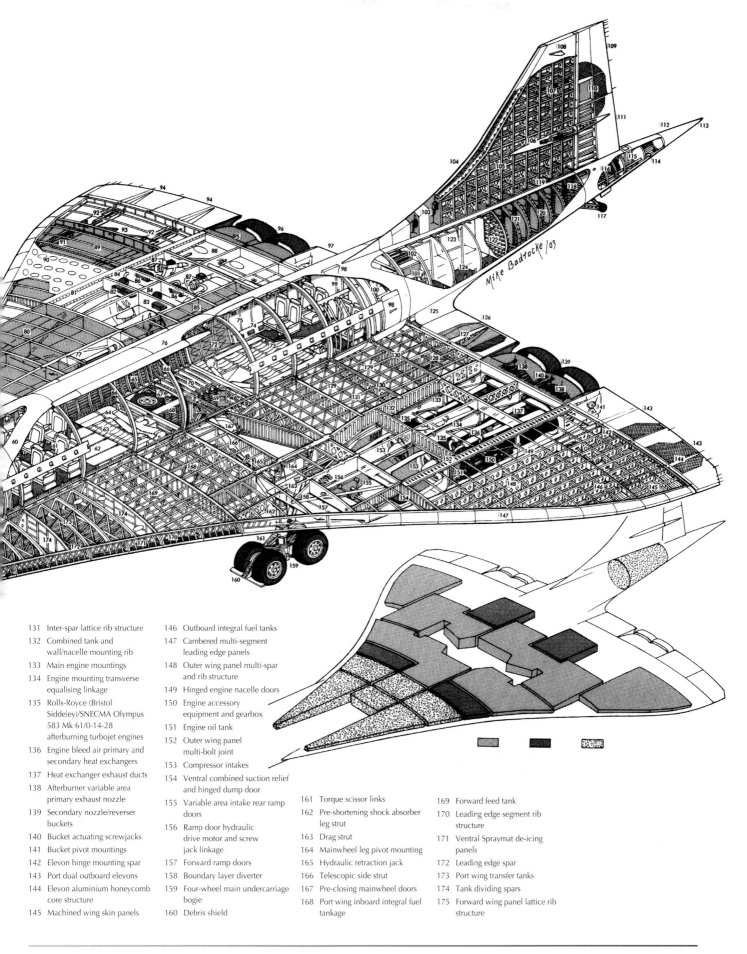

131 Inter-spar lattice rib structure
132 Combined tank and wall/nacelle mounting rib
133 Main engine mountings
134 Engine mounting transverse equalising linkage
135 Rolls-Royce (Bristol Siddeley)/SNECMA Olympus 583 Mk 61/0-14-28 afterburning turbojet engines
136 Engine bleed air primary and secondary heat exchangers
137 Heat exchanger exhaust ducts
138 Afterburner variable area primary exhaust nozzle
139 Secondary nozzle/reverser buckets
140 Bucket actuating screwjacks
141 Bucket pivot mountings
142 Elevon hinge mounting spar
143 Port dual outboard elevons
144 Elevon aluminium honeycomb core structure
145 Machined wing skin panels

146 Outboard integral fuel tanks
147 Cambered multi-segment leading edge panels
148 Outer wing panel multi-spar and rib structure
149 Hinged engine nacelle doors
150 Engine accessory equipment and gearbox
151 Engine oil tank
152 Outer wing panel multi-bolt joint
153 Compressor intakes
154 Ventral combined suction relief and hinged dump door
155 Variable area intake rear ramp doors
156 Ramp door hydraulic drive motor and screw jack linkage
157 Forward ramp doors
158 Boundary layer diverter
159 Four-wheel main undercarriage bogie
160 Debris shield

161 Torque scissor links
162 Pre-shortening shock absorber leg strut
163 Drag strut
164 Mainwheel leg pivot mounting
165 Hydraulic retraction jack
166 Telescopic side strut
167 Pre-closing mainwheel doors
168 Port wing inboard integral fuel tankage

169 Forward feed tank
170 Leading edge segment rib structure
171 Ventral Spraymat de-icing panels
172 Leading edge spar
173 Port wing transfer tanks
174 Tank dividing spars
175 Forward wing panel lattice rib structure

CONCORDE UTILISATION DETAILS

CONCORDES BUILT IN UNITED KINGDOM

Aircraft Registration	No.	1st Flight	Hours	Landings	Supersonic Flights	Reference Flights	Retired to
G-BSST	002	April 1969	836	438	196	—	Yeovilton
G-AXDN	101	December 1971	633	273	168	—	Duxford
G-BDDG	202	February 1974	1,435	—	633	—	Brooklands
G-BOAC	204	February 1975	22,260	7,729	6,760	6,890	Manchester
G-BOAA	206	November 1975	22,769	8,064	6,842	7,086	Edinburgh
G-BOAB	208	May 1976	22,297	7,810	6,688	6,974	Heathrow
G-BOAD	210	August 1976	23,397	8,406	7,010	7,294	New York
G-BOAE	212	March 1977	23,376	8,383	7,003	7,357	Barbados
G-BOAG[1]	214	April 1978	16,239	5,633	5,066	5,078	Seattle
G-BOAF[2]	216	April 1979	18,257	6,045	5,639	5,555	Filton
BA total			**148,595**	**52,070**	**45,008**		

Prototype: 002, Pre-production: 101, Production: 202–216, only 204–216 flown commercially.

[1] formerly G-BFKW, [2] formerly G-BFKX

The rig at Farnborough achieved 20,000 equivalent supersonic cycles. Originally measured in supersonic cycles the new yardstick, for BA, became 'reference flights' which included a compensation for light take-off weights (see chapter 'In Service'). However there was no finite limit to the life of the airframe.

CONCORDES BUILT IN FRANCE

Aircraft Registration	No.	1st flight	Hours	Landings	Supersonic Flights	Reference Flights	Retired to
F-WTSS	001	March 69	812	397	249	—	Le Bourget
F-WTSA	102	Jan 73	657	314	189	—	Orly
F-WTSB	201	December 73	910	423	247	—	Toulouse
F-BTSC[3]	203	Jan 75	11,989	4,481	3,978	—	Crashed
F-BVFA	205	October 75	17,824	6,780	5,504	—	Washington
F-BVFB	207	March 76	14,771	5,473	4,791	—	Sinsheim
F-BVFC	209	July 76	14,322	4,358	4,200	—	Toulouse
F-BVFD	211	February 77	5,814	1,929	1,807	—	Broken up
F-BTSD[4]	213	June 78	12,974	5,135	3,672	—	Le Bourget
F-BVFF[5]	215	December 78	12,421	4,259	3,734	—	CDG
AF total			**90,115**	**32,415**	**27,686**		
Grand total			238,710	84,485	72,694		

Prototype: 001, Pre-production: 102, Production: 201–215, only 203–215 flown commercially.

[3] formerly F-WTSC, [4] formerly F-WJAM, [5] formerly F-WJAN

BRITISH AIRWAYS CONCORDE CREWS 1976–2003

Captains

RPW Allen
R Anderson
A Baillie*
M Bannister*
S Bates
KP Barton
RJL Boas
JAD Bradshaw
DA Brister
NA Britton
LD Brodie*
AL Budd
JD Butterley
JW Burton
BJ Calvert
JL Chorley
JD Cook
RV Dixon
PM Douglas*
PRW Duffey
JD Eames
DA Edmondson
DH Ellis
M Emmett
V Gunton
JA Harkness
T Henderson
JW Hirst
PW Horton
JC Hutchinson
KD Leney
HJ Linfield
WD Lowe
P Mallinson
AJ Massie
CB McMahon
HC McMullen
IC McNeilly
AR Meadows

A Mills*
RA Mills
CJC Morley
CC Morris
GF Mussett
K Myers
C E Norris*
M O'Connor
B Oliver (CAA)
T Orchard
R Owen*
DJM Rendall
J Reid* (CAA)
E Reynolds
MA Riley
S Robertson
M Robinson
DG Ross
DC Rowland
L Scott
RS Smith
D Studd
A Thompson*
BGT Titchener
NV Todd
J Smith
BO Walpole
SD Wand
R Weidner*
R Westray*
JR White*
G Williams* (CAA)
D Woodley

Total 72

First Officers

AB Atkinson
RP Babbé
P Bandall
A Barnwell*

J Bedforth*
P Benn*
P Benson
AJ Bird
S Bohill-Smith
ME Boyle
RJS Burchell
MW Burke
C Burrough
D Byass*
AD Cobley
AR Darke
J Downey
L Evans*
J Graham
CD Green
P Griffin*
B Harmer
AI Heald
BR Holland
J Hornby
JG Huson
B Irven
M Jealous*
DA MacDonald-
   Lawson
A Macfarlane
DG Mitchell
G Mitchell
EER Murton
J Napier*
BR Oliver
CJD Orlebar
D Payne
JH Phillingham
J Phillips
R Pike
WJ Piper
AF Quartly
NS Rendall

JM Reynolds
R Reynolds
CA Robey
ML Robson
J Rooney
RA Routledge
PT Sinclair
WI Smith
K Snell
K Strocchi
RJ Taylor
J Tye
M Walden*
MHJ Watson
D Whitton
K Williams
MR Withey
MR Young
AJ Yule

Total 62

Engineer Officers

K Appleton
SL Bolton
GS Bowden
RC Bricknell
K Brotherhood*
AA Brown
WJ Brown
PC Carrigan*
A Chalmers*
TF Caster
C Coltman
L Cooper*
M Cooper
TB Dewis
W Dobbs
FW Duffy
R Eades
PE Egginton

T Evans
A Everett
IR Fellowes-Freeman
P Finlay
SG Floyd
F Ford
J Goatham
M Harrison
W Hazelby*
M Hollyer*
WG Hornby
D Hoyle*
S Hull
D Jackman
WD Johnston
WA Johnstone
RA Jones
IV Kirby
JE Lidiard
PA Ling
DA Macdonald
R Maher*
PJ Newman
T Norcott*
PJ Phillips
A Price
TJ Quarrey
I Radford
JA Rodger
IF Smith
J Stanbridge
D Tracey
G Tullier
A Walker*
RN Webb
NC West
AF Winstanley
J Wood*
R Woodcock*

Total 57

This is a list of all the BA crew members that have ever been qualified to operate Concorde. In 1980 there was a maximum of 28 qualified crews, since then the average was around 20, dropping to 14 in the mid-1990s and around eight in November 2001. Those qualified to operate Concorde as of April 2003 are marked with an asterisk. A 'crew' consists of Captain, First Officer and Engineer Officer.

Captains

F Andreani
G Arondel
B Biras
D Blay
J Boye
M Brulant
M Butel
G Caillat
G Campion
C Catania
P Chanoine Martiel
J-L Chatelain
J-M Chauve
J Chauvin
E Chemel
E Chillaud
P Conte
R Contresty
P Debets
C Delorme
R Demeester
R Doguet
J Doublait
A Duchange
P Dudal
G Duguet
M Dupont
W Dupont
M Ferry
HG Fournier
G Fremont
A Gely
P Gilles
P Girard
P Gourguechon
H Grandjean
F Hertert
C Hetru
G Jacob
P Lacoux
J Lalanne
C Leclerc
G Le Gales
M Le Guillou
J-P Le Moel
JC Leopoldes
JC Lesieur
M Lortsch
R Machavoine
D Malbrand
D Manchon
C Marty
A Massoc
J-F Michel
J Mims
J Moron
Y Pecresse
P Plisson
M Pouligny
J Prunin
Quilichini
J Ramon
M Rio
B Robert
J Robin
J Rossignol
F Rude
R Sagory
J Schwartz
G Tardieu
R Terry
A Verhulst
F Vicens
C Voog

Total 74

First Officers

F Adibi
B Bachelet
G Barras
A Bataillou
A Bernigaud
P Bonnot
E Celerier
D Chambrier
A Colloc
D Compagnon
D Costes
A Croise
P d'Haussy
PA Decamps
P Delangle
JC Delorme
B Depouez
M Doumax
A Dubreucque
C Durieux
G Duval
D Fady
G Fourtier
A Geoffry
JF Gibouin
P Giron
P Grange
J Guegan
AJ Holbecq
J Jaillet
P Lapersonne
J Le Breton
D Le Chaton
J Marcot
B Marchand
M Martin
G Mauroy
G Metais
JM Milliere
S Morisset
P Neutre
R Ortiz
R Othnin Girard
JM Peloffy
F Pradon
JM Proust
R Puyperoux
L Ravera
G Ringenbach
G Rogon
E Tonnot
M Tronche
R Vacchiani
B Vialle

Total 54

Engineer Officers

JP Aubry
A Babot
G Barbaroux
P Baty
H Benard
R Beral
C Billerey
A Blanc
R Bonzi
G Brugeroux
R Buisson
V Cappoen
Cazin
D Casari
G Clement
B Collette
S Coulombel
JP Crouzet
G Cucchiaro
A Czmal
J-P Desserprit
M Detienne
M Diou
JY Dronne
M Drouard
P Dubourg
R Duguet
R Escuyer
L Faviez
L Frot
G Jardinaud
J Jarrousse
Lafaye
J-J Lajarge
A Lavillaureix
P Le Berre
C Lebrun
J Ledoux
J Lombart
F Marquis
JL Masselin
B Menoret
H Michaut
G Moustier
G Pellerin
A Piccinini
H Pivet
Y Pluchon
C Poulain
H Ranty
A Roger
L Schwaller
M Suaud
J Thomas
S Vallet
D Vasseur

Total 56

CONCORDE CHRONOLOGY

1956

November 5: Supersonic Transport Aircraft Committee (STAC) established.

1959

March 9: STAC recommends design studies of two supersonic airliners (Mach 1.2 and Mach 2).

1959–61

French and British SST feasibility and design studies initiated.

1961

First Anglo-French discussions on possible SST collaboration. *June–July:* First BAC/Sud discussions in Paris and Weybridge.

1963

Preliminary design presented to airlines. *January 13:* President de Gaulle uses name 'Concorde'. *June 3:* Concorde sales option issued by Pan American Airlines. *June:* BOAC and Air France sign Concorde sales options. *June 5:* US supersonic transport programme announced.

1964

Projected medium-range version abandoned: design of long-range version enlarged. *May:* Enlarged Concorde design announced. *July:* Olympus 593F first run, Bristol.

1965

April: Metal cut for Concorde prototypes. *May:* Pre-production Concorde design announced.

1966

Manufacturer and airline engineering liaison established. *April:* Final assembly of prototype 001 begins, Toulouse. *June:* Concorde flight simulator commissioned. *August:* Assembly of prototype 002 begins at Filton. *September:* First flight of Avro Vulcan testbed with Olympus 593 engine. *October:* Olympus 593 achieves 35,190lb thrust on test at Bristol. *December:* Fuselage section delivered to RAE, Farnborough for fatigue testing.

1967

Design for pre-production aircraft revised to reduce drag: rear fuselage extended, new nose/visor. *February:* Concorde interior mock-up presented to airlines at Filton. *April:* Complete Olympus 593 engine first test-run in high-altitude chamber, Saclay. *December 11:* 001 rolled out at Toulouse. British partner adopts 'Concorde' spelling.

1968

February: UK government announces £125 million loan for Concorde production. *August:* 001 taxi trials, Toulouse. *September:* rolled out at Toulouse. British aircraft design redefined. *December:* Olympus 593 ground testing reaches 5,000 hours. *December 31:* First flight of Tupolev Tu-144 prototype.

1969

March 2: First flight of 001, Toulouse. *March:* Government authority given for construction of nine airframes (two ground test airframes, two prototypes, two pre-production aircraft and three production aircraft). *April 9:* First flight of 002, Filton (to Fairford). *June:* First public appearance of both prototypes at Paris Air Show. *October 1:* 001 achieves Mach 1. *November 8:* Airline pilots fly 001. *December:* Authority given for construction of three more production aircraft.

1970

March 25: 002 achieves Mach 1. *September 1:* 002 appears at SBAC show, Farnborough; lands at Heathrow. *November 4:* 001 achieves Mach 2. *November 12:* 002 achieves Mach 2.

1971

January: 100th supersonic flight. *March 24:* Congress stops US supersonic transport programme. *April:* Authority given for four more production aircraft. *May 25:* 001 appears at Paris Air Show; flies to Dakar. *June:* Total Concorde flight time reaches 500 hours. *August:* 100th Mach 2 flight. *September 20:* First pre-production Concorde, 01, rolled out at Filton. *December 17:* First flight of 01, Filton.

1972

February 12: 01 exceeds Mach 1. *April 13:* Production aircraft 11–16 authorised. *April 22–23:* 002 appears at Hanover Air Show. *May 25:* BOAC announces intention to order five Concordes. *June:* 002 sales demonstration tour of Middle East and Australia. *July 24:* China signs preliminary purchase agreement for two Concordes. *July 28:* BOAC orders five Concordes, Air France orders four. *August 28:* China signs preliminary purchase agreement for a third Concorde. *September 28:* Concorde 02 rolls out at Toulouse. *October 5:* Iran Air signs preliminary purchase agreement for two Concordes plus option on a third.

1973

January 10: First flight of 02, Toulouse. *January 22–February 24:* 002 completes 'hot and high' trials at Johannesburg. *January 31:* Pan American and TWA decide not to take up their Concorde options. *June 3:* Production Tu-144 crashes during Paris Air Show. *June 30:* 001 flight to Fort Lamy (Chad) gives 80-minute scientific observations of solar eclipse. *September 18:* 02 leaves Paris for first US visit for opening of Dallas/Fort Worth Airport. *September 26:* Breaks Washington–Paris record (3hrs 33mins) on return flight. *October 19:* 001 retired to French Air Museum at Le Bourgeat. *December 6:* First flight of 201, first production Concorde.

1974

February 17–19: 02 completes low-temperature trials at Fairbanks, Alaska. *February 13:* First flight of 202, second production Concorde (G-BBDG). *June 25:* Concorde static test specimen at CEAT tested to destruction. *July 19:* Initial production programme of 16 aircraft agreed. *August:* Middle East demonstration flights by 202. *September 12:* Flight testing total reaches 3,000 hours. *October 20–28:* American Pacific coast demonstration tour. *October:* Supersonic flight total reaches 1,000 hours.

1975

January 31: First flight of 203. *February 11:* Completion of passenger emergency evacuation certification trials. *February:* Tropical icing trials. *February 27:* First flight of 204. *February:* Certification trials at Madrid. *May 28:* Special category C of A for 203 awarded; registration changed to F-BTSC, start of 'endurance' flying by this aircraft (completed August 2). *May:* Static display and flying programme at Paris Air Show. *June 30:* CAA special category C of A for 204 awarded. *October 9:* Concorde receives French C of A. *October 14:* British Airways and Air France open reservations for Concorde scheduled services. *October 25:* First flight by 205. *November 5:* First flight by 206. *November 13:* Final Environmental Impact Assessment published by FAA. *December 5:* Concorde receives British C of A. *December 19:* Air France receives its first Concorde (205).

1976

January 5: Concorde public hearing held by US Secretary of Transportation. *January 6:* Air France receives 203. *January 15:* BA receives its first Concorde (206). *January 21:* Airline service begins, London–Bahrain (BA, 206) and Paris–Rio via Dakar (Air France 205). *February 4:* Concorde services to New York and Washington for 16 months trial period approved by US Secretary

of Transportation. *February 13:* BA receives its second Concorde (204). *March 4:* 002 retired to RN Air Station, Yeovilton. *March 6:* First flight of 207. *April 8:* Air France receives 207. *April 9:* Air France service extended to Caracas via Santa Maria, Azores. *May 18:* First flight of 208. *May 20:* 02 retired to Orly Airport, Paris. *May 24:* Transatlantic flights to Washington begin, from both Paris and London. *July 9:* First flight of 209. *August 13:* Air France receives 209. *August 25:* First flight of 210. *September 30:* BA receives 208, its third Concorde. *November 30:* Fairford flight test base closed. *December 6:* BA receives 210. *December 8:* 203 returned to Aérospatiele.

1977

February 10: First flight of 211. *March 17:* First flight of 212. *March 26:* Air France receives 211. *July 20:* BA receives 212. *August 20:* 01 retired to Duxford, under care of Imperial War Museum. *October 19:* Proving flight to New York begin. *October 26:* Singapore Airlines/BA agreement on London–Singapore flights via Bahrain announced. *November 2:* HM the Queen and HRH Prince Philip return from Barbados on Concorde. *November 22:* Services to New York begin, from Paris and London. *December 9:* Service from London to Singapore via Bahrain begins.

1978

Prolonged talks on Malaysian Concorde ban. *April 21:* First flight of 214. *June 26:* First flight of 213. *August 10:* BA carries its 100,000th Concorde passenger. *September 18:* Air France receives 213. *September 20:* Air France opens service Paris to Mexico City via Washington. *December 26:* First flight of 215.

1979

January 9: US type certificate awarded. *January 12:* Braniff subsonic service between Washington and Dallas/Fort Worth inaugurated. *January 24:* BA/Singapore Airlines extension Bahrain–Singapore resumed. *February 22:* British Government announces BA to write off Concorde purchase cost: Government to receive 80% of operating surpluses. *April 20:* First flight of last production Concorde (216), Filton. *September 21:* British and French governments announce unsold aircraft and support engines to be placed with BA and Air France. *December 16:* BA Concorde flies London–New York in 2hrs 59mins 36secs.

1980

February 6: BA receives 214. *June 1:* Braniff ceases Dallas/Fort Worth service. *June 13:* BA receives 216. *October 23:* Air France receives 215. *November 1:* Singapore services discontinued.

1981

January 21: Five years in airline service; 50,000 hours, 15,800 flights, 700,000 passengers. *January–February:* Evidence on Concorde presented to Commons Industry and Trade Committee. *April 14:* Report on Concorde published by Commons Industry and Trade Committee expresses dissatisfaction with cost figures and urges efforts to make sure that costs are shared equally with France. Government's reply in July describes committee's criticisms of forecasts as 'unwarranted'. *September 11:* Anglo-French 'summit' meeting; British and French governments commission joint study on future of Concorde. *October 29:* British and French meet in London to discuss Concorde. Three options proposed (1) cancellation, (2) a phased rundown (3) indefinite continuation. *December 2:* British Government review of relative costs presented to Parliament by Department of Industry. *December 9:* Department of Industry ministers and officials give evidence to Commons Trade and Industry Committee.

1982

March 31: Air France services to Caracas and Rio discontinued. *February:* Commons Trade and Industry Committee reaffirms dissatisfaction with cost aspects. *May 1:* Formation of Concorde Division within BA, responsible for profitability of Concorde operations. *May 6:* British and French ministers meet in Paris to discuss Concorde (cost reduction, officials' report, cost-sharing). *August:* Ian Sproat, minister responsible, writes to BA Chairman Sir John King stating that the government will cease to fund Concorde's British manufacturers (Rolls-Royce and British Aerospace). *October:* Sir John King replies to the Sproat letter saying that BA will examine the possibility of Concorde funding the support costs out of revenue. *November:* Group set up within BA to examine support costs.

1983

January 1: Fastest transatlantic crossing west-to-east, New York–London in 2hrs 56mins. *April 13:* SFO Christopher Orlebar is technical consultant to BBC TV programme on Concorde in QED series, broadcast today.

1984

March 27: Concorde's inaugural flight to Miami via Washington. Henceforth a thrice-weekly service will operate. *March 31:* After 18 months of negotiations, the British Government's involvement in the Concorde project becomes minimal, with BA becoming responsible for funding Concorde's British manufacturers. *September 11:* Distance record Washington–Nice by G-BOAB

(4,565 statute miles). *November 16:* Concorde (G-BOAB) inaugural charter to Seattle from London via New York.

1985

February 13: First commercial service London–Sydney by Concorde under charter, establishing a record time of 17hrs 3mins. *March 28:* Concorde under a commercial charter establishes the record between London and Cape Town of 8hrs 8mins. *April 25:* New livery unveiled by Concorde G-BOAG (214) returning into service. This aircraft had been out of service for a long period, with much of its equipment having been removed for use in the other Concordes. *May 11:* Concorde special charter inaugural to Pittsburgh. *December 19:* Highest recorded ground speed by Concorde to date, achieved by G-BOAC in commercial service, 1,292kts (1,488mph). throughout 1985 Concorde inaugurates several routes within Europe for publicity purposes and to destinations within the Americas, in conjunction with the Cunard shipping line.

1986

January 21: Concorde celebrates 10 years of commercial operations. *April 5:* First Concorde charter to New Zealand, viewing Halley's Comet over the Indian Ocean. *July 11:* Prime Minister Margaret Thatcher makes her first supersonic flight on Concorde from London to Vancouver to visit EXPO 86. She is presented with a copy of *The Concorde Story*. *November 8–23:* Concorde's first round-the-world charter. Special edition of *The Concorde Story* commissioned by John Player to mark the occasion.

1987

January: Concorde spearheads advertising campaign for the privatisation of BA. *March 7:* Second round-the-world trip, organised by Goodwood Travel. *September 6:* Captain Brian Walpole sets new transatlantic speed record – just 95mins between Hopedale, Newfoundland, and the north-west Irish coast. *October 5:* BA Concorde carries millionth transatlantic scheduled passenger – Patrick Mannix of Reuters. *November 22:* Concorde celebrates tenth anniversary of operations into New York JFK. Richard Noble, land speed record-breaker, sets new record by crossing the Atlantic three times in one day – on Concorde. *December 12:* Concorde commences new winter schedule, once weekly to Barbados for the winter season.

1988

December 7: Concorde G-BOAA makes record New York–Heathrow flight, 2hrs 55mins 15secs. *April:* G-BOAA is first

Concorde to undergo 12,000 flying hours check. Structure pronounced sound for service well into 21st century. Inaugural Concorde twice weekly to Dallas for summer period until August. *June:* Captain Walpole awarded OBE in Queen's birthday honours. December: Concorde resumes Barbados service for winter season. Captain Walpole retires from BA.

1989

February 7: Noel Edmonds features Captain John Hutchinson on BBC programme about Concorde, aviation and travel. *March 2:* Aerospatiale host party to celebrate 20th anniversary of Concorde's first flight at Toulouse. *March:* HM the Queen flies to Barbados on Concorde. *April 1:* Concorde leaves Heathrow on supersonic circumnavigation charter, to cover 38,215 miles. *April:* Concorde loses part of rudder between Auckland and Sydney, circumnavigation continues after repair. *November 19:* Channel 4 TV programme *Faster Than a Speeding Bullet* presented by Captain Christopher Orlebar features Concorde in the history and future of supersonic flight.

1990

January 21: Concorde's 14th anniversary of commercial service. January: Goodwood Travel notches up 37,153 charter passengers on Concorde. *April 6:* Prince Michael of Kent presents awards to Concorde engineers. *April 14:* Captain Norman Britton takes New York–London record with 2hr 54min 30secs flight. *May 5:* BAe and Aerospatiale unveil plans for Concorde successor; BA will take 'more than usual interest'. *June 1:* Concorde flies Honolulu–Hong Kong in 6hrs 30mins (subsonic time 13hrs 10mins). *June 6:* Celebration of 50th anniversary of Battle of Britain. Captain WD 'Jock' Lowe flies Concorde in formation with Spitfire over the White Cliffs of Dover. *August 10:* Concorde assists at Gatwick Airport's Diamond Jubilee. *September 7:* Concorde model installed at Heathrow's main roundabout. *September 14:* Concorde appears at Farnborough Air Show. *November 30:* Concorde twice-weekly winter service to Barbados resumes.

1991

January 27: Miami services cease due to Gulf War, two months earlier than originally envisioned. *January:* Second Concorde suffers rudder delamination. *April 4:* Captain Hutchinson celebrates 15th anniversary of Washington's supersonic service. *May 5:* Captain Jock Lowe becomes Director of Flight Crew BA. *May 14:* HM the Queen and HRH Prince Philip fly in Concorde to Washington to start official tour of USA, Captain Lowe in command. *June 7:* New computerised Concorde flight planning

system introduced. *July 12:* Concorde 01 at Duxford repainted. *July 19:* General Sir Peter de la Billière awards tickets to Gulf War veterans. *July 26:* Concorde celebrates opening of Birmingham's new airport terminal. *October 18:* IMF bankers to Bangkok on Concorde. *November 1:* Captain Lowe hosts Jackie Mann, former Beirut hostage. December: Captains Hutchinson and Musset each break a record, Acapulco to Honolulu via Las Vegas and Bali to Mombasa via Colombo.

1992

January 22: Mary Goldring, former fierce critic, looks forward to a Concorde successor in her TV programme *Goldring's Audit*. *February 21:* Captain Hutchinson, nearing retirement, broadcasts on BBC World Service *A Personal View*. *March:* Concorde to visit Nigeria and South Africa. *March 27:* Neil Kinnock, electioneering for the Labour Party, visits BA Concorde. *April:* Tories win General Election. *July 17:* Captains Horton, Hutchinson and Musset – Concorde speed record-holders – to Buckingham Palace garden party. July: Former BOAC chairman Sir Basil Smallpeice dies. *October 12–13:* Concorde Spirit Tours sets westbound circumnavigation record of 32hrs 49mins 3secs in F-BTSD from Lisbon. *November 11:* Concorde pulled 440yds by RAF team for Children in Need. *December:* Captain David Rowlands becomes Concorde Flight Technical Manager (FTM). *December 18:* Concorde to Barbados again for winter with one extension to Mexico at end of March 1993.

1993

January 29: Lord King, pictured with Concorde model tucked under arm, is honoured for Outstanding Contribution to International Aviation during his chairmanship of BA; he is to become President of BA for life. *March 26:* Senior First Officer Barbara Harmer of BA becomes first lady Concorde pilot. *May 21:* Concorde G-BOAF to become first of seven to have £1m internal and external refurbishment. *July 2:* Captain Rowland (FTM) to also be Concorde Business Manager. *August 27:* Concorde human tow to raise money for BA charity 'Dreamflight'. *September 12:* Captain Orlebar, former Concorde first officer, consultant for *Your Flight*, Channel 4 documentary on air traffic control. *October 10:* Special flight Concorde to Jeddah. *November 12:* Concorde with Captain Morris at Athens celebrating 60th anniversary of Greek association. *November 19:* G-BOAA receives new £1m upper rudder, solution to previous rudder problem. *December 3:* 'Fastest Show on Earth' – the Bee Gees perform aboard Concorde on a flight around the Bay of Biscay for Children in Need, organised by Concorde SEO Bill Brown.

1994

January 7: Concorde BA 001 Heathrow–New York first flight with new cabin crew uniform. *February 4:* Student from Southampton awarded Sword of Honour on Concorde's flight deck. *March 2:* 25th anniversary of Concorde 001's first flight from Toulouse is celebrated. March: Concorde dinner at which Sir George Edwards and Brian Trubshaw are guest speakers. *June:* Captain Rowland resigns as combined FTM and Business Manager. *August:* Cracks found in Concorde wing in a rear spar web structure, non-primary and easily repaired. *August:* Three outer window panes cracked at Mach 2 and 57,000ft – no loss of pressurisation, with inner ply windows withstanding pressure. Replacement outer windows now double ply as well. *October:* Richard Branson announces possibility of Virgin Atlantic leasing Air France Concorde plus crews. *October:* Concorde exhaust analysed in flight near New Zealand. *October:* Concorde ceases services to Washington. This will save 300 sectors per year. *November:* Concorde takes BA delegates on official visit to India.

1995

January: Captain Mike Bannister becomes Flight Technical Manager Concorde. *March:* Concorde dinner at RAF Club. *April 2:* Prime Minister John Major flies to Washington on Concorde 'to trump US President Clinton'. *May:* Disagreement between Aerospatiale and partners over future SST (a Mach 2 350-seater); Germany and Britain favour subsonic 600+ seater; Terrazoni (GM of Aerospatiale) doubts whether the required $20bn could be found but sees a market, especially for reducing the 14hr subsonic flights over the Pacific. *May 29:* Sir Archibald Russell CBE FRS, British designer of Concorde, dies aged 90. *August:* Captain Mike Bannister takes over Concorde training responsibilities as Flight Manager Concorde. *August 15–16:* F-BTSD sets eastbound circumnavigation record from JFK, with Concorde Spirit Tours. *September:* NASA tests a hypersonic wing configuration in wind tunnel at Langley, Virginia. *September 25:* Victorious Ryder Cup team returns via Dublin with trophy on board Concorde. *September:* NASA prepares to make supersonic choices at the HSR (High-Speed Research) programme. *September:* Olympus 593 engines clock up half a million hours of supersonic flight; meanwhile Rolls-Royce and SNECMA continue to work on the Mid-Tandem Fan non-reheat engine. *September 21:* RAeS lecture by Ian Gray of BAe: 'Is the Concorde successor getting closer? – Yes!' *October:* British and French airworthiness authorities will decide in 1996 on necessary modifications to extend Concorde's life up to 8,500 Reference Flights. *October:* NASA commences flight tests on Supersonic

Laminar Flow Control. *November:* Sir Colin Marshall to become non-executive chairman of BA and Robert Ayling to become chief executive. *December:* Captain Jock Lowe to stand down as Director of Flight Crew.

1996

January: Collision avoidance radar, essential for flight over USA, fitted successfully to Concorde, overcoming problems associated with overheating of aerial at supersonic speeds. *February:* British and French airworthiness authorities rule that only six changes to maintenance schedules (so no modifications) will be required for Life Extension Programme, due to excellent state of airframes. *February 7:* Captain Leslie Scott breaks New York–London record, with a 2hr 52min 59sec flight. *March:* NASA completes synthetic vision landing system tests using video and infra-red sensors, thus future SST would not need droop nose. *March 17:* A modified Tu-144L is rolled out for six months of joint Russian-US (NASA) research. *March:* Proposals for the X-33 reusable single-stage to orbit will be submitted to NASA by May. *April 2:* Air France Concorde painted blue to advertise Pepsi-Cola's new can. *April 15:* NASA unveils supersonic concept aircraft, Mach 2.4, 9,200km (5,200nm) range, 310 seats, noise, stage 3 rules less 3dB. *May 20:* BA announces record profits for year 1995/96 of £580m. *May 27:* Concorde G-BOAG in for 'major' and refurbishment. *June 2:* Concorde and the Red Arrows in formation flypast for 50th anniversary of Heathrow Airport. *August 8:* Sir Frank Whittle, inventor of the jet engine, dies aged 89. *August 22:* Channel 4 TV documentary suggests that 1973 Paris Air Show Tu-144 crash was caused by Soviet pilot's violent manoeuvre to avoid 'spying' French Mirage – story denied due to Soviet/French collusion. *September 3:* Julian Amery (Lord Amery of Lustleigh), former Minister of Aviation and signatory of the 1962 'Concorde Treaty' with France, dies. *September 7:* Concorde appears at Farnborough Air Show. *October 10:* G-BOAG's (216) 'major' check completed – last BA Concorde 'major' before the millennium. *October 1:* Boeing predicts that future SST would create 'turmoil' by 'siphoning off' First, Business and Full Fare Economy equating to 66% of the revenue on a conventional service. *November 29:* Tu-144LL flew again from Zhukovsky for a series of 32 joint Russian–US research flights. *December:* TBB – former Concorde hangars at Heathrow dismantled; purpose built facility with 'tail end docking' introduced.

1997

January 21: Concorde celebrates 21 years in service. *February 11:* 30 million telephone calls made for £10 New York return Concorde tickets (normally over £5,000) issued to celebrate

10th anniversary of BA privatisation. *April:* Air France (AF) reactivates Concorde F-BVFB from storage, cost put at $5.2 million, as others in the fleet approach their 12,000 hour 'major' checks. *May 15:* BA Concorde services to Barbados to run from June to August. *May 18:* 'Gas shroud' proposal from California University could cut engine noise. *May 28:* BA Concorde fares up 15% to £5,596 LHR–JFK return, passenger numbers unaffected thanks to soaring UK economy. AF Concorde fares up only 8%. *June 20:* Concorde Alpha Foxtrot displaying the new BA 'Union Flag World Image' livery conveys Prime Minister Blair to the Denver Summit. *July:* Aleksandr Pukhov, Tupolev Chief Designer, announces that fewer than 32 missions will complete flight testing of the Tu-144LL (joint Russian–US project) – five flights versus 15 sufficed for subsonic testing. The remaining ten flights will start in September. Maximum speed achieved to date is Mach 2 – $30,000 of fuel is consumed on each flight. *July:* 'Life Extension' now confirmed up to 8,500 reference cycles with capability to exceed this after review in 2004–2005 – no modifications required, merely changes to routine inspections. *August:* MDC flies 6% scale model of the inherently unstable 'Blended-Wing-Body' design for very efficient subsonic flight – 800 seats over 7,000nm. A Concorde successor might benefit from such technology. If the BWB is successful the economic criteria for the Concorde successor might be more difficult to meet. *August:* David Gower, former cricketer, to present Concorde on forthcoming BBC documentary *The Air Show. August 27:* Japan announces intention to build Concorde successor, 300 seats with Mach 2.4 (1,600mph) cruise. *September:* Brad Faxon, of the US Ryder Cup team, achieves an 8-mile putt on Concorde at 1330mph (lasting 23 seconds) during a record breaking, three hour 25 minute flight from New York to Malaga captained by Jock Lowe. *September 23:* Brian Trubshaw, Jock Lowe, Jim O'Sullivan (Chief BA Concorde Engineer) and Bob McKinlay present 21 Years of Concorde with BA at the Royal Aeronautical Society London. *November:* United Space Alliance, responsible for Space Shuttle servicing, visit BA to study techniques of Concorde. *November 20:* 'Lottery Live' shows winner Mark Gardiner arriving in Barbados on Concorde.

1998

January 21: Concorde 22 years in service. *February 4:* Iraqi arms inspection crisis, PM Tony Blair to Washington in Concorde. *April:* Tu-144LL flight tests ends, achieving 39 hours over 19 flights. *April:* General Electric studies non-variable cycle engine for Dassault's proposed Supersonic Business twinjet 'Falcon SST' – 67kN dry and 98kN with reheat, M1.8 cruise.

May 20: 75th anniversary of Farnborough celebrated with Concorde visit – Capt Lowe i/c. *May 25:* Part of left middle elevon breaks off one hour into flight to JFK, safe return to London, no long term problem envisaged. *July 1:* England World Cup team, defeated by Argentina, returns in Concorde. *July 7:* Sir Robert Lickley, chief Designer Fairey Delta 2 (FD2) dies aged 86. *September 12:* Concorde at Farnborough for 50th anniversary of air show. *October:* Tu-144 set for a further eight test flights with NASA. *November 21:* Capt Dave Studd and crew celebrate 21st anniversary of Concorde operations into New York.

1999

January 17: Sunday Times prediction: a Supersonic 8 seater business jet will fly by 2006. *January 18:* Tories criticise Labour for using Concorde to Washington. *February:* NASA abandons HSR – no near term market apparent. *February:* France and Russia to study 'Scramjet' for Mach 2.5-12. *March 2:* 30th anniversary of Concorde 001's first flight. *March 4:* US threatens Concorde's JFK landing rights in Trade War. *March 5:* Brian Trubshaw speaks at Concorde dinner, RAF club London. *March 17:* Dassault shelves SSBJ due to engine problems – noise. *March 23:* AF Concorde uneconomic; but BA bullish. *April 9:* 30th anniversary of 002's maiden flight, Goodwood Special Flight to Filton. *May:* Concorde landing rights threat lifted as EU climbs down over hushkitted older jets question. *May 24:* Manchester United take Concorde to Barcelona for European Cup vs Bayern Munich. Two days later, United win 2-1. *June:* NASA rolls out X-34 low cost orbital technology demonstrator. *June 10:* A very apologetic Harvey Weinstein, producer of *Shakespeare in Love* and *The English Patient,* fined £200 for smoking in a Concorde lavatory. *June 30:* Executive Jet (pioneer of 'Fractional Ownership') joins Lockheed Martin and Gulfstream in pressing NASA to fund Business Jet X-plane. *July 1:* Concorde GBOAE, with Red Arrows, opens the Scottish Parliament Edinburgh with fly-past. *August 8:* Concorde £14 million cabin refurbishment by Sir Terence Conran announced. *August 11:* Concorde special flight tracks final solar eclipse of the millennium. *September 10:* Retired Concorde Captain David Ross dies. *September 18:* James Doyle 76, alias Jimmy Cameron, admits to being 'Ace' – the BAC engineer who divulged Concorde secrets to Tupolev. *December 31:* Concorde, heard but not seen (low cloud), overflies The London Eye (BA's Ferris Wheel) as part of the Millennium Celebrations.

2000

January: Lockheed Martin and Gulfstream reveal ways for minimising sonic booms in the QSAT (Quiet Supersonic

Aircraft Technology) programme. A 4000nm, Mach 1.8 'Demonstrator' might fly in five years. *January:* £14 million cabin refurbishment commences. *April 12:* Lionel Haworth, Aeroengineer who developed R-R Olympus engine for Concorde dies aged 87. *May:* Infrared technology – to detect atmospheric turbulence ahead of 2-D inlet for QSAT – to be tested on a Lockheed SR-71. *July:* Cracks reported in some Concorde wings, spar 72. *July 25:* AF Concorde 203 crashes near Paris CDG. Number 2 engine fire causing Number 1 failure as well thought to be cause. 109 souls on board killed, five in hotel at Gonesse die too. Capt Christian Marty was i/c of this German charter flight to New York. BA and AF suspend Concorde flights. *July 26:* BA resume Concorde flights; AF states: no resumption until cause has been established. *July 29:* Fuel leak caused by tyre burst and subsequent fire ignited by reheat, now thought to be the cause. *July 30:* Due to possible fuel smell on a BA flight to New York, Concorde diverts to Gander. *August 7:* 40cm piece of metal found on runway, after V1 point, could have contributed to Concorde crash; main gear water deflector also implicated; parallels to AF Concorde incident at Washington Dulles 14 June 1979 noted. *August 15:* BA grounds its Concorde fleet due to imminent withdrawal of C of A. Capt Les Brodie is possibly the last BA Concorde pilot to fly from JFK. *August 16:* DGAC and CAA withdraw Concorde's C of A. *August 31:* Preliminary report published. *September:* Edgard Chillaud returns 'stranded' AF Concorde from JFK to CDG. *September 15:* Leslie Appleton, designer of Fairey Delta 2 (FD2) dies aged 86. *November 7:* Judicial report states that the metal strip (with tyre rubber traces) found on the runway fell from a Continental Airways DC10. *November 17:* Remains of Concorde F-BTSC being reconstructed at Hangar 12 Dugny – near Le Bourget. French Magistrate Xavier Salot states 'we know it was not a bomb'. The two engines that failed are examined at the Centre for Propulsion Studies, Sanglay (a suburb of Paris) *December:* Approval for modifications given. *December:* First Michelin NZG tyre to Dunlop for fitting to Concorde main wheel. CEAT Toulouse perform dynamometer tests on new combination.

2001

January: Damage tests carried out on Michelin NZG tyre Almeria Spain. *January 5:* Interim report on Concorde crash published by BEA. *January 14:* AF Concorde flies CDG to Istres, near to Marseille, to investigate behaviour of fuel leaks in airflow under the wing. Cause of fire still not understood. *January 16:* £17 million to be spent on structural modifications at same time as cabin refurbishments. *February:* AF Pres J-C Spinetta announces that AF is as keen as BA to return

Concorde to service. *March 24:* Brian Trubshaw, British Concorde test pilot dies aged 77. *May:* AF Concorde F-BTSD returns from Istres after tyre and brake tests. *May 12:* Alexei Tupolev, designer of Tu144 ('Concordski'), dies aged 76. *July 5:* Richard Wiggs, founder of Anti-Concorde dies, aged 72. *July 17:* G-BOAF completes 'very successful' supersonic test flight LHR to Brize Norton via Atlantic. *August:* BA pilot retraining, Concorde at Shannon, Ireland. *September 5:* Concorde's C of A restored. *September 11:* Following a 'dress rehearsal' flight, the Concorde return to service team hear of the attacks on the Pentagon and World Trade Center. *September 20:* Brian Trubshaw memorial service St Clement Dane's Church, The Strand, London. *October 22:* G-BOAF returns to JFK for an 'assessment flight', Capt Les Brodie i/c –'last out first in' (see August 00). *November 7:* BA and AF resume services to New York. Mayor Rudi Giuliani greets Concorde's clientele. BA Capt Mike Bannister, AF Capt Edgard Chillaud. *November 7:* PM Tony Blair takes Concorde (dubbed 'Blair Force One') to Washington, for talks with President GW Bush concerning Afghan war and Middle East. *December 1:* Concorde resumes weekly Barbados services. *December:* Mayor Giuliani arrives by Concorde to receive honorary Knighthood from Queen Elizabeth. *December 16:* Promotional tickets for New Year period at one third normal price sell out in three minutes.

2002

January 21: 26th anniversary of Concorde service. *January 27:* Fourth modified Concorde in each airline ready for testing. *March 24:* Jeremy Clarkson writes amusing and supportive article about 'the big white dart' in the *Sunday Times*. *April 16:* Barbados winter flights season comes to a successful end. *April 16:* Eastbound service rescheduled to depart JFK at 8.30am and arrive LHR 5.25pm. *April 18:* Gordon Brown, British Chancellor, flies Concorde to US for G7 meeting. 'AD' suffers engine surge en route and is 30 minutes late. *June 4:* Concorde AD Capt Mike Bannister in Jubilee fly-past over Buckingham Palace with Red Arrows in formation. *July 14:* Japanese NAL rocket launched 1/10th scale research craft fails. *July 15:* AE LHR–JFK returns to LHR, due to engine surge. *August 1:* Scramjet tested at Mach 7 at Woomera, rocket launched, very short burn. *August 3:* Summer Barbados schedules recommence. Minor mod to rev thruster allows higher take-off weight, so direct both ways (London and Barbados) becomes possible, without a refuelling stop. *November 27:* 'AE' suffers lower rudder delamination en route LHR–JFK – first such incident since 1998, fifth overall. *December 4:* Story of rudder hits press, but passengers are not put off. With only three modified aircraft available, the later

JFK–LHR schedule is reactivated. *December 14:* BA and AF receive award for returning Concorde to service from l'Academie Aeronautique Francaise. Capt Les Brodie accepts on BA's behalf.

2003

January 21: 27th anniversary of Concorde service. *January:* Boeing cancels 'Sonic Cruiser' – due to 'lack of airline interest'. *February 19:* AF Concorde diverts to Halifax due to engine shut down and fuel leak. *February 26:* Rumours of Concorde services terminating due to reduced revenue are denied by BA. *February 27:* AF Concorde suffers rudder delamination CDG to JFK, no danger say experts. *March 2:* Sir George Edwards, former Chairman of BAC dies, aged 94. *March 6:* Iraq/UN crisis: UK Foreign Secretary Jack Straw, interviewed in flight to JFK supersonic. *March 7:* RAF Club Piccadilly: Tony Benn is Guest Speaker at Concorde Dinner. *April 1:* BA Concorde LHR–JFK departure changed to evening to benefit from US market, corporate Concorde bookings down, engineering costs up. Mandatory anti-terrorist cockpit door costs $300,000. AF suffering from unofficial US boycott due Chirac's anti Iraq war stance. *April 10:* Iraq: Saddam Husein's statue is felled. SARS virus hits Far East. World airline revenue falls. BA announces cessation of Concorde operation as of 31 October 2003 and AF as of 31 May 2003. Run down to cost £80 million. Concorde bookings soar to record levels. *April 14:* Richard Branson's Virgin Atlantic bids for Concorde – 'a national asset'; CAA sceptical, BA negative. *April 30:* Channel 4 TV Lord Marshall, BA Chairman dismisses Branson's bid with: 'the manufacturers will not support Concorde after 31 October 03', 'did you tell them to say that?', 'No.' *April 30:* £7 million bid for Concorde nose cone. *May 1:* Lord Marshall in *The Times*, announces that BA would have flown Concorde 'beyond October' if AF and BA had been united. In the event Airbus wanted to redeploy its workforce more profitably than supporting Concorde. Former Concorde chief pilot, Jock Lowe, says 1962 treaty, if invoked, would have forced France to support Concorde. *May 6:* Branson bid to save Concorde cannot expect government aid. *May 8:* BA Chief Executive Rod Eddington on BBC 1's Question Time, 'the Great plane will be retired'. *May 8:* AF Concorde F-BTSD top rudder failure. AF denies that such failures caused their premature wish to ground Concorde; although there were

appalling load factors resulting from French opposition to US war in Iraq 'exorbitant costs were to blame'. *May 31:* AF ceases Concorde operations. *June 15:* Concorde F-BTSD arrives Le Bourget. *June 27:* Concorde F-BVFC arrives Toulouse, the final flight of the French Concordes. *July 9:* Toulouse Concorde under judicial seal, effect of missing undercarriage 'spacer' in 2000 crash to be scrutinised. *July 21:* Concorde 'FB' is transported by barge and 'low loader' to arrive at Sinsheim Museum Germany for reassembly and display. *August 7:* Concorde take-off weight reduced due to UK record high temperatures (35°C [95°F]) causing refuelling stop at Gander en route JFK. *August 20:* 'FB' nearing reassembly at Sinsheim. *August 30:* G-BOAC leaves Barbados for the last time. *September 6:* Airbus appears unwilling to support a 'heritage' Concorde, flying only on special occasions. *September 17:* JFK–LHR, engine surge causes crew to decelerate early. Range subsonic is less so 'AG' is diverted to Cardiff. *October 12:* New NZG tyre on Concorde deflates 'quietly' after hitting 'foreign object' during take-off. *October 24:* Crowds jam Heathrow to greet final three Concorde landings. *November 5:* Concorde 'AG' breaks record JFK to Seattle, three hours 55 minutes 12 seconds – Canadian government approves supersonic flight over tundra. *November 6:* Winston Bray, former BOAC director involved with acquisition of Concorde, dies at 93. *November 10:* Concorde 'AD' arrives in JFK. *November 16:* Christie's Paris auctions Concorde memorabilia. Nose cone sold for £287,500 versus expected £10,000. *November 17:* Concorde 'AE' leaves LHR for Barbados. *November 25:* Concorde 'AD' carried on a barge to rest alongside USS *Intrepid*. *November 26:* Concorde G-BOAF final flight LHR to Filton supersonic over Bay of Biscay. *December 1:* Bonhams auctions BA Concorde memorabilia. Nose cone sells for £320,000, Machmeter, £28,000; £750,000 raised for disabled children's charity.

2004

January: QinetiQ marketing 'Tarsier' – a system for debris detection capable of scanning a runway six times per minute. *January 21:* Concorde G-BOAB handed over to Heathrow on 28th anniversary of entry into service. Will be exhibited as part of Terminal 5 complex. 'AB' was chartered and flown by author on 50th anniversary of Schneider Trophy, 12 September 1981.

BA CONCORDE DESTINATIONS FROM LONDON (HEATHROW) AIRPORT, AS OF 2003

24 October 2003 BA ceased scheduled services
Final scheduled services: Barbados (non stop): once weekly (or more) during winter season and August; To New York: daily 10.30am–09.25am, latterly 6.30pm–5.25pm; from New York: daily 9.30am–6.00pm
Earlier scheduled services: Bahrain: thrice weekly; Dallas Fort Worth (via Washington with BA, then Braniff): thrice weekly; Miami (via Washington): thrice weekly; Singapore (via Bahrain): thrice weekly; Toronto: various schedules over summer months; Washington Dulles: thrice weekly

Diversions:
Atlantic City
Bangor
Boston
Gander
Gatwick
Halifax
Lajes
Montreal
Newark
Shannon
Windsor Locks

Charter Destinations of both Air France & BA. (AF) = exclusive to Air France (up to 1998).

Africa:
Abidjan
Agadir (AF)
Antanarivo (AF)
Aswan
Bamako (AF)
Bujumbura (AF)
Cairo
Capetown
Cartagene (AF)
Casablanca
Conakry (AF)
Dakar
Djerba (AF)
Douala (AF)
Djibouti (AF)
Harare
Johannesburg

Kigali (AF)
Kinshasa (AF)
Kilimanjaro
Lagos (AF)
Lome (AF)
Libreville (AF)
Lusaka (AF)
Luxor
Marrakesh
Mauritius (AF)
Moroni (AF)
Mombasa
Nairobi
Ouagadougou (AF)
Reunion (AF)
Robertsville (Monrovia)
Seychelles (AF)
Tangier (AF)
Tozeur (AF)
Tunis (AF)
Yamoussoukro (AF)

The Americas:
Abbotsford
Acapulco
Albany
Andrews AFB (AF)
Anchorage (AF)
Antigua
Aruba
Asheville
Atlanta
Atlantic City
Austin (Texas)
Baltimore
Bangor
Barbados

Barreirinhas (AF)
Battle Creek
Bermuda
Boston
Brasilia (AF)
Bogota (AF)
Buenos Aires (AF)
Buffalo
Calgary
Caracas
Cayenne (AF)
Charleston
Chicago (AF)
Cincinnati
Cleveland
Colorado Springs
Columbus
Dayton
Denver
Detroit
Edmonton
Fort-de-France (AF)
Fort Lauderdale
Fort Myers
Goose Bay
Grand Cayman
Hampton (AF)
Harrisburg
Hartford/Springfield
Havana (AF)
Honolulu
Houston Ellington
Houston Intnl
Iguassu (AF)
Indianapolis
Jackson
Jacksonville
Kailua-Kona

Kansas City
Kingston
Las Vegas
Lexington
Lima
Little Rock
Llanbedr
Lubbock
Mexico City
Miami
Midland-Odessa
Moncton (AF)
Montego Bay
Montevideo (AF)
Montreal
Nashville
Nassau
Newburg Stewart
New Orleans
Newport
New York
Oakland
Oklahoma City
Omaha
Ontario (USA)
Orlando
Oshkosh
Ottawa
Papeete (AF)
Philadelphia
Phoenix
Pittsburgh
Pointe-a-Pitre (AF)
Port-au-Prince (AF)
Portland
Port of Spain
Providence
Puerto Rico

Quebec (AF)
Raleigh
Recife (AF)
Regina (AF)
Reno (AF)
Richmond
Rio de Janeiro
Rochester
Rockford
Sacramento (AF)
St Louis
St Lucia
Saint-Martin (AF)
Salt Lake City
San Antonio
San Diego (AF)
San Francisco (AF)
San Juan (AF)
Santa Maria
Santiago
Santo Domingo (AF)
Sao Paulo (AF)
Schenectady
Seattle Boeing Field
Seattle Tacoma
Springfield
Syracuse
Tahiti
Tampa
Tampa Macdill
Toronto
Trinidad
Tucson
Val d'Or (AF)
Vancouver
Waco
Washington Dulles
Washington Andrews

Wichita
Wilmington
Windsor Locks
Windsor Ontario

Australasia:
Auckland
Brisbane
Christchurch
Darwin (AF)
Easter Island (AF)
Fiji
Hao (AF)
Jakarta (AF)
Learmonth
Mururoa (AF)
Noumea (AF)
Perth
Sydney

Europe and Middle East:
Aarhus (AF)
Aalborg (AF)
Abu Dhabi (AF)
Albacete (AF)
Amman
Amsterdam
Ancona
Ankara (AF)
Aqaba
Athens
Badajoz (AF)
Baikonour (AF)
Bahrain
Barcelona
Basel
Bastia (AF)
Beauvais (AF)
Beirut (AF)
Bergen
Berlin
Biarritz (AF)

Billund
Bologna
Bordeaux
Bratislava
Brest (AF)
Brussels
Budapest
Cambrai (AF)
Charleroi (AF)
Chateauroux
Clermont-Ferrand (AF)
Cologne
Copenhagen
Dijon (AF)
Dhahran
Dubai (AF)
Dublin
Epinal (AF)
Faro
Frankfurt (AF)
Geneva
Graz
Grenada
Grenoble (AF)
Gothenburg (AF)
Hamburg (AF)
Hanover
Haifa (AF)
Helsinki
Ibiza (AF)
Istanbul
Ivalo (AF)
Jeddah
Kangerlussuaq (Sondrestrom)
Keflavik
Klagenfurt (AF)
Kish (AF)
Kuwait
Lajes (AF)
Lanzarote (AF)
Larnaca

Las Palmas (AF)
Leipzig
Liege (AF)
Lille (AF)
Linz
Lisbon
Luxembourg
Lyon (AF)
Madrid
Malaga
Malta (AF)
Marseille (AF)
Metz (AF)
Milan Linate
Milan Malpensa
Montpellier (AF)
Moscow
Mulhouse (AF)
Munich
Munster
Muscat (AF)
Nantes (AF)
Nice
Novossibirsk (AF)
Nuremherg (AF)
Oporto
Oslo
Ostend (AF)
Patina
Paris CDG
Paris Orly
Paris Le Bourget
Pescara
Pisa
Poitiers
Prague
Reins (AF)
Riyadh
Rome Fiumicino
Rovaniemi
Sana'a (AF)
S-J de Compestella (AF)

St Petersburg
Salzburg
Seville
Stavanger
Stockholm
Strasbourg
Tarbes (AF)
Tenerife
Tel Aviv
Toulouse
Tours
Turin
Turku (AF)
Vasteras
Valladolid (AF)
Venice
Vichy (AF)
Vienna
Warsaw

Far East:
Bali
Bangkok
Bangui (AF)
Beijing
Bombay
Calcutta
Chiang Mai (AF)
Colombo
Dacca (AF)
Delhi
Denpasar (AF)
Djakarta
Guam
Hong Kong
Islamabad (AF)
Kathmandu (AF)
Kuala Lumpur
Madras
Nagasaki (AF)
Osaka/Kansai (AF)
Singapore
Tashkent

Tehran (AF)
Tianjin (AF)
Tokyo (AF)

United Kingdom:
Aberdeen
Belfast
Birmingham
Boscombe Down
Bournemouth
Brize Norton
Cardiff
Coltishall
Derby (AF)
East Midlands
Edinburgh
Exeter
Fairford
Farnborough
Filton
Finningley
Gatwick
Glasgow
Hatfield
Humberside
Kinloss
Leeds
Leuchars
Liverpool
Luton
Macrihanish
Manchester
Manston
Mildenhall
Newcastle
Prestwick
St Mawgan
Stansted
Teeside
Yeovilton

Air France Concorde destinations from Charles de Gaulle Airport (Paris)
31 May 2003 AF ceased scheduled services

Final scheduled services: Paris–New York: five days per week 10.30am–8.25am; New York–Paris: five days per week 8.00am–5.45pm
Earlier Scheduled Services: Caracas (via Santa Maria); Mexico (via Washington); Rio de Janeiro (via Dakar); Washington Dulles

BIBLIOGRAPHY

An Introduction to the Slender, Delta Transport, BAC/ Aerospatiale (1975).

Beaty, David, The Water Jump, Secker & Warburg (1976).

British Airways and Concorde Finances, Report of the Review Group, Department of Trade and Industry (February, 1984).

Burnet, Charles, Three Centuries to Concorde, MEP (1979).

Calder, PH & Gupta, PC, Future SST Engines, SAE (1975).

Calvert, Brian, Flying Concorde, Airlife (1981).

Concorde, Second Report from the Industry and Trade Committee, Session 1980–81.

Flight International, various articles.

Gupta, PC, Rolls-Royce Ltd., Aero Div., Bristol, Advanced Olympus for Next Generation Supersonic Transport Aircraft, SAE (1981).

History of Aviation, New English Library (1972).

Hooker, Stanley, Not Much of an Engineer, Airlife (1984).

Jane's All the World's Aircraft, various editions.

Knight, Geoffrey, Concorde, The Inside Story, Weidenfeld & Nicholson (1976).

Leyman, CS, BAe, Bristol, ' After Concorde What Next?', presentation at AIAA annual meeting, Washington (1985).

Morpurgo, JE, Barnes Wallis, Longman (1972).

Owen, Kenneth, Concorde, New Shape in the Sky, Jane's (1982).

Peterson, Richard H, & Driver, Cornelius, 'Advanced Supersonic Transport Status', presentation at AIAA annual meeting, Washington (1985).

Reed, Arthur, Britain's Aircraft Industry, What Went Right? What Went Wrong?, Dent (1973).

Sinton, Darrol, The Anatomy of the Aeroplane, Granada (1966, reprinted 1980).

Smallpeice, Basil, Of Comets and Queens, Airlife (1981).

Whittle, Frank, Jet, the Story of a Pioneer, Frederick Muller (1953).

ILLUSTRATIONS ACKNOWLEDGEMENTS

Together with those individually credited, photographs and diagrams were kindly supplied by the following; Aeroplane; Aerospace Publishing Ltd; Peter A. Bisset; The Boeing Company; Braniff International; British Aerospace; British Tourist Authority; Avions. Marcel Dassault; Flight International; International Air Radio Ltd; Adrian Meredith Photography; Metro Dade County; M. L. Nathan; National Aeronautics and Space Adminstration; Nigel Paige; Christine Quick Photography Ltd; RAF Museum; Rolls-Royce; Jeppesen Sanderson Inc; Skye Aerial Photographers; Michael Turner.

INDEX

References to illustrations are shown in **bold**.

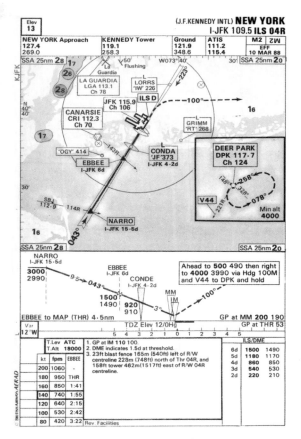

(J.F.KENNEDY INTL) NEW YORK
I-JFK 109.5 ILS 04R

Elev 13

NEW YORK Approach	KENNEDY Tower	Ground	ATIS	M2 ZW
127.4	119.1	121.9	111.2	EFF
269.0	258.3	348.6	115.4	10 MAR 88

SSA 25nm 28 SSA 25nm 2o

LA GUARDIA LGA 113.1 Ch 78
LORRS 'IW' 226
JFK 115.9 Ch 106 ILS D
CANARSIE CRI 112.3 Ch 70
GRIMM 'RT' 268
'OGY' 414
CONDA 'JF'373 I-JFK 4·2d
EBBEE I-JFK 6d
DEER PARK DPK 117·7 Ch 124
NARRO I-JFK 15·5d
V44 Min alt 4000

Ahead to **500** 490 then right to **4000** 3990 via Hdg 100M and V44 to DPK and hold

NARRO I-JFK 15·5d **3000** 2990
EBBEE I-JFK 6d
CONDE I-JFK 4·2d **1500** 1490 **920** 910
MM IM **200** 190

EBBEE to MAP (THR) 4·5nm
TDZ Elev 12/0Hg
GP at MM **200** 190
GP at THR 53

Var 12°W

T.Lev	ATC
T.Alt	18000

kt	fpm	EBBEE
200	1060	-
180	950	THR
160	850	1:41
140	740	1:55
120	640	2:15
100	530	2:42
80	420	3:22

1. GP at IM 110 100.
2. DME indicates 1.5d at threshold.
3. 23ft blast fence 165m (540ft) left of R/W centreline 228m (748ft) north of Thr 04R, and 158ft tower 482m (1517ft) east of R/W 04R centreline.

ILS/DME

6d	1500	1490
5d	1180	1170
4d	860	850
3d	540	530
2d	220	210

Rev. Facilities

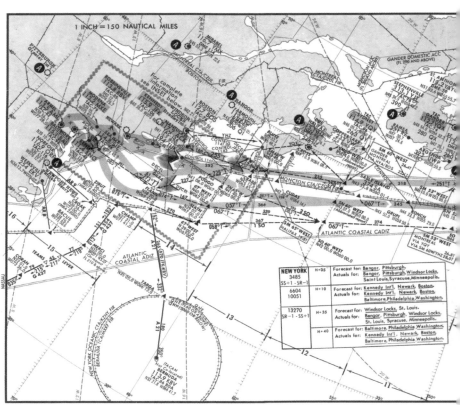

1 INCH = 150 NAUTICAL MILES

NEW YORK		
SS+1 - SR-1	3485	Forecast for: Bangor, Pittsburgh. Actuals for: Bangor, Pittsburgh, Windsor Locks, Saint Louis, Syracuse, Minneapolis.
H+10	6604 10051	Forecast for: Kennedy Int'l, Newark, Boston. Actuals for: Kennedy Int'l, Newark, Boston, Baltimore, Philadelphia, Washington.
SR-1 - SS+1	13270	Forecast for: Windsor Locks, St. Louis. Actuals for: Bangor, Pittsburgh, Windsor Locks, St. Louis, Syracuse, Minneapolis.
H+40		Forecast for: Baltimore, Philadelphia, Washington. Actuals for: Kennedy Int'l, Newark, Boston, Baltimore, Philadelphia, Washington.

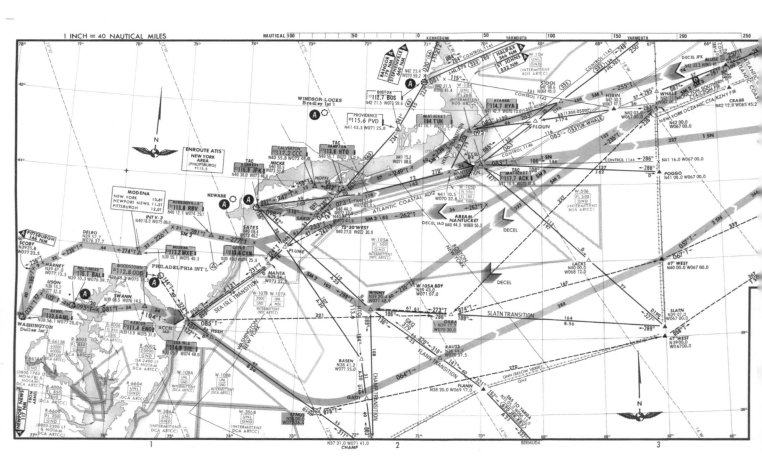

1 INCH = 40 NAUTICAL MILES